# Introduction to
# Business Information Systems

W9-AZT-603

Dear Mark,

with a particular
dedication for
you – and
many thanks for
your support

Wolffing

Springer
*Berlin*
*Heidelberg*
*New York*
*Hong Kong*
*London*
*Milan*
*Paris*
*Tokyo*

Rolf T. Wigand · Peter Mertens
Freimut Bodendorf · Wolfgang König
Arnold Picot · Matthias Schumann

# Introduction to Business Information Systems

With 79 Figures

 Springer

Professor Dr. Rolf T. Wigand
Maulden-Entergy Chair and Distinguished
Professor of Information Science
and Management
Department of Information Science
CyberCollege
University of Arkansas at Little Rock
2801 South University Avenue
Little Rock, AR 72204-1099, USA
rtwigand@ualr.edu

Professor Dr. Dr. h.c. mult.
Peter Mertens
Bereich Wirtschaftsinformatik I
Friedrich-Alexander-Universität
Erlangen-Nürnberg
Lange Gasse 20
90403 Nürnberg
Germany
mertens@wiso.uni-erlangen.de

Professor Dr. Freimut Bodendorf
Bereich Wirtschaftsinformatik II
Friedrich-Alexander-Universität
Erlangen-Nürnberg
Lange Gasse 20
90403 Nürnberg
Germany
bodendorf@wiso.uni-erlangen.de

Professor Dr. Wolfgang König
Johann-Wolfgang-Goethe-Universität
Frankfurt
Institut für Wirtschaftsinformatik
Mertonstraße 17
60054 Frankfurt
Germany
koenig@wiwi.uni-frankfurt.de

Professor Dr. Dres. h.c. Arnold Picot
Ludwig-Maximilians-Universität
München
Institut für Organisation
Ludwigstraße 28
80539 München
Germany
picot@bwl.uni-muenchen.de

Professor Dr. Matthias Schumann
Georg-August-Universität
Göttingen
Institut für Wirtschaftsinformatik
Platz der Göttinger Sieben 5
37073 Göttingen
Germany
mschuma1@gwdg.de

ISBN 3-540-00336-3 Springer-Verlag Berlin Heidelberg New York

Cataloging-in-Publication Data applied for
A catalog record for this book is available from the Library of Congress.
Bibliographic information published by Die Deutsche Bibliothek
Die Deutsche Bibliothek lists this publication in the Deutsche Nationalbibliografie; detailed bibliographic
data is available in the Internet at <http://dnb.ddb.de>.

Springer-Verlag Berlin Heidelberg New York
a member of BertelsmannSpringer Science+Business Media GmbH

http://www.springer.de

© Springer-Verlag Berlin · Heidelberg 2003
Printed in Germany

Softcover-Design: Erich Kirchner, Heidelberg

SPIN 10907832      42/3130-5 4 3 2 1 0 – Printed on acid-free paper

# Preface

Business information systems is the discipline which investigates, categorizes and evaluates methods to systematically develop and deploy information systems in companies and organizations. Its importance increases in the evolving information society. Many areas of modern life and work are supported by the deployment of information systems. Thus the necessity arises to convey essential features of business information systems to readers in educational settings at various levels. This book is to support such learning efforts.

In contrast to most other introductory books, the presentation is *consistently oriented towards integrated application systems*. Subject matters such as the technology of computers, programming, as well as data storage are emphasized a little less in their relative importance, especially since the authors set themselves a strict page limit.

The authors assume basic knowledge on the student's side in the area of computers and computer networks gained from the university's PC laboratory or from private PC usage at home. Initially essential basic knowledge about hardware and software is provided. Starting with the PC, the features and characteristics of other computer classes are developed and the fundamentals of networks, especially the Internet, are presented. As students progress in their business studies, the present textbook shows how processes in firms are supported by information systems. It provides concepts that are used in modern application systems. Moreover, the integrated perspective of these applications advances the thinking, conceptualization and imagination underlying operational processes. For example, towards the end of the student's business studies the relations among various management areas and functions (sales, production, accounting, etc.) are easier to comprehend and to observe.

To characterize the ever-increasing importance of business information systems in organizations as well as for the global economy we have made available numerous practical examples. Consequently, we offer extensive discussions about the Internet and phenomena in relation to it, like supply chain management, electronic procurement, networked databases and electronic commerce. The authors are fully aware that it is difficult to stay on top of the fast-paced developments in the field of business information systems. We believe though that our joint effort offers a balanced view of these developments within the available space.

The authors owe Rolf Wigand a great thank for the initial translation from German into English. Moreover, the following individuals offered valuable assistance in the preparation of this book: Dipl.-Wirtsch.-Inf. Thomas Franke (chapters 1 and 4, as well as sections 5.1, 5.2.7.4, 5.2.8.2 and 5.4), Dipl.-Kfm. Markus Fricke (chapter 2), Dipl.-Math. Andreas Jahn and Dipl.-Wi.-Ing. Ulrich M. Löwer (chapter 3), Dr. Susanne Robra-Bissantz and Dipl.-Kfm. Bernd Weiser (sections 5.2, 5.3, as well as 4.3.1), Dipl.-Wirtsch.-Inf. Joachim Rawolle (chapter 6) and Prof. Dr. Thomas Hess (chapter 7). Dipl.-Volksw. Roman Beck coordinated all tasks with great enthusiasm. Prof John Kanet, University of Dayton, helped us to find specific terms in the field of production and materials management.

We would like to thank our readers in advance for any form of feedback while using this book.

*The Authors, March 2003*

# Contents

# 1 The Subject of Business Information Systems

Business information systems or just information systems (IS) deal with the conceptualization, development, introduction, maintenance and utilization of systems for computer-assisted information processing within companies and enterprise-wide networks. Central components of such information systems are a firm's application systems (AS). They assist the user in the firm to accomplish tasks.

A long-term goal for information systems is to automate anywhere in the enterprise and in the economy where a task may be executed by a computer system at least as well as by a person, i.e. with regard to quality, costs, etc. (*meaningful full automation* [Mertens 95]). When such automation cannot be achieved, information systems nevertheless are at least to support specialists and executives as effectively as possible.

Beyond this, logical networking of business processes (within and among enterprises) may serve as a tool to leverage potentials. Logical networking comes about through the physical networking of application systems.

## 1.1 Examples of Business Application Systems

The following examples were chosen to offer a first impression of the multitude of AS:

1.  The salesperson of a truck manufacturer visits a shipping firm and takes a portable notebook along. He enters a description of the business and specifically information about the shipping volume. The personal computer (PC) finds a suitable truck with the necessary accessories, calculates the price, estimates the operating costs of the vehicle and makes available an appropriate financing plan of the purchase. After the customer decides to make this purchase, largely based on the carefully thought-through offer, the PC transmits the order to a computer at the headquarters of the truck manufacturer.

2.  An application system in the plant helps to plan the production of the ordered trucks for each individual calendar week and also takes care of the ordering of materials to be delivered by suppliers, such as tires and seats.

3. During the manufacturing of the trucks an AS controls automated drilling machines, turning lathes and other machining tools, welding robots, as well as those devices for quality control. Moreover, such AS coordinate the delivery of parts, as well as the storing of the produced parts.

4. A computer-supported planning system generates the forecast of truck sales for the next few years, as well as the demand for manufacturing capacity and the needed capital.

5. In a business within the pharmaceutical industry an application system controls the delivery of raw materials to the chemical reactors and regulates parameters, such as pressure and temperature. Next, the AS guides the manufactured substance to an automated machine that, in turn, presses pills and makes sure that the right pills, packing foils, instruction sheets and cartons arrive at the right point in time at the wrapping machine.

6. A business in the airline and space industry deploys a multimedia system for the training and development of workers. They are thus enabled to take courses about innovative technologies from nearly any location worldwide. These are attractively enhanced by graphics, animated pictures, short videos and sounds. Written and spoken text alternate in the presentation.

7. At the cash register of a supermarket an AS captures information of all items sold via the bar code. It searches the corresponding item description and prices in the data base of the store's computer, prints the customer's receipt and deducts each sold item from its corresponding inventory. Moreover, such an AS may also link each purchase to an individual customer and his/her buying habits, if the customer used a loyalty card.

8. In a shipping firm a routing and trip planning program configures the best possible allocation of goods to be shipped to specific travel routes and trucks, prints loading documents and instructions for the delivery personnel, as well as trip and delivery orders for the drivers.

9. A mail order firm makes it possible for its customers to locate on the Internet at any time of day where the customer's package is at the present point in time and which distribution points have been passed already.

10. In a bank the AS manages the accounts of customers. It records payments and money transfers received. Moreover, the system records payouts made and scheduled transfers of funds, calculates interest and issues account statements.

11. An insurance company uses an application system for the calculations of risks based on the accepted contracts and informs management accordingly.

12. A mortgage company operates an intelligent mortgage AS that links customer application data with the customer's credit data and employ-ment/earning history and makes a decision within minutes instead of 30 days, as was customary in the not too distant past.

13. In a city building department an AS manages a building permit applica-tion via e-mail on the computer monitors of those reviewing the applica-tion, seeks evaluations and reminds participants of open decisions that are overdue.

14. A travel agency uses the computer to show empty seats on a particular flight and, simultaneously, to make reservations for a seat and a hotel, as well as to rent a car in the destination city. Then the computer issues the invoice and books the trip. In addition, the travel agency can find out readily last-minute deals on the Internet.

15. In a university an AS captures the data about new students, prints stu-dent ID cards and IDs to use the library and to get public transportation passes at reduced prices. Moreover, the system generates student statis-tics to be used by the university administration.

Our examples here do not just represent certain businesses or industries, but also completely different task types of application systems. In the case of managing accounts in the bank (example 10), a management process is re-constructed and streamlined through the use of the information system. Such systems we call *administrative systems*. In example 8 (shipping business) we highlighted the partially or completely automated scheduling of the process, i.e. a *disposition system*. Administrative and disposition systems may be grouped under the label *operating systems*. The application system in exam-ple 4 supports the planning of production capacities in the car manufacturing industry, thus it utilizes a *planning system*. The application system mentioned in example 11 permits the management of the insurance company to control its risk situation of its business. It follows that this is a *control system*.

With *administration systems* one strives to achieve efficiency within exist-ing processes, e.g., in order to give bank employees more contact time for attending to customers (case 10). *Disposition systems* aim for improved deci-sions. For example, in case 8 round-trips are to be identified for which the product "tons x miles" is smaller than of those figured out by an individual manually (e.g., the dispatcher or driver). At the same time the shipping agency is also strengthening its competitive position, as it can offer lower prices. *Planning systems* make sure that more reliable data are available for the planning process and that more alternatives may be considered and calcu-lated. *Control systems* grab the attention of executives and experts by demon-strating noteworthy data constellations where specific analyses and adjust-ments may be necessary. In case 11, e.g., the board of directors of the insur-ance company will affect additional reinsurances when the information sys-tems show significant imbalances, i.e., unreasonable high risk. Planning and control systems are often closely interrelated, hence one often refers to *P&C systems*. Operating systems generally support staff more at the lower and middle, whereas planning and control systems are more likely to support individuals at the middle and higher levels of the organization hierarchy (cf., fig. 1.1/1).

Application systems do not operate as isolated systems. For example, the AS mentioned in case 2 gets its data about the sold trucks from the AS of case 1. The third AS makes sure that the needed parts from case 2 are indeed manufactured. Also the AS in case 4 utilizes the information about the truck sales (case 1) for its planning purposes. In an ideal situation the information system coordinates all events in the "Customer Order Processing" of the vehicle manufacturer. Application systems should be carefully coordinated and be in tune with each other. They are to operate based on common data-sharing. This does not necessarily mean that these data need to be stored in the same location. The basis for this is the concept of *integrated information processing*. In our case we are concerned with *intraorganizational integrated information processing* of the truck manufacturer.

*Fig 1.1/1        Application inside the Organization Pyramid*

In supermarkets (case 7) it is not necessarily a requirement that the customer has cash or checks available. At the registers the customer typically may pay by using debit or credit cards. The purchase amount will be transmitted via a dedicated telecommunications line to the computer of the buyer's bank and will there be immediately deducted from his/her bank account (if using a debit card). If the buyer uses a credit card, the amount will be debited against his account and will appear on the monthly statement listing all charges for that month. This process requires that the information systems of several firms have to be coordinated and be able to operate among each other. Such a solution is referred to as an *interorganizational integrated information system*.

## 1.2 Information as a Factor in Production

Information may be differentiated from the data controlled by, e.g., an information system and by the needs of a respective user. The term information, consequently, refers to the *meaning* and *intended purpose* of the data. It follows then that the information content that a business manager may deduce from the data "80 percent of the sold products were developed less than two years ago" is different than the same content deduced by a supplier. For the business manager this means that his/her business must place a high value on flexible and market-oriented product development as a competence, whereas the supplier is likely to interpret the content to mean a threat for his/her market position vis-à-vis frequent product changes.

Information is undoubtedly a dominating factor in our modern economy and society. We refer to an *information society* and mean by that the successor of traditional, machine-based material production. Information and knowledge have always played an important role for societal development. Production based on the division of labor has always demanded coordination based on information and communication. Decisive for the development of an information society, however, are the changes in degree of information and communication in the economy [Picot et al. 01, pp. 21 et sqq.; Wigand et al. 97]. Increased coordination activities among ever more-specialized actors trigger an increasing demand for technical information and communication equipment.

Information and communication costs, determined by their relative degree of intra- and interorganizational coordination, i.e. so-called *transaction costs*, increase steadily within the economy. Based on a 1986 study [Wallis/North 86, p. 121 et sqq.] an increase in the share of transaction costs in the GDP of the U. S. was reported from 25 to 55% for the period of 1870 to 1970. It is safe to assume that today this share in modern economies is exceeding 60%. It follows that information has become a production factor within single, but also within entire economies. The increase in the division of labor and specialization in general, the alignment of enterprises, product lines and markets resulting in ever-more specialized segments of the value chain offer a high potential for productivity increases. On the other hand, they also demand more efficient coordination and fine-tuning. In these efforts information and communication technology makes a decisive contribution by utilizing innovative organizational solutions. This increase in information and communication intensity observable within the economy can only then offer an advantage, if the productivity gains realized are higher than the additional coordination costs generated. It is precisely here where we realize the strategic importance of information and communication technologies: It lowers the costs of coordination and makes new forms of the division of labor and alignment possible.

The content of this text should be seen in the light of the increasing importance of information as a production factor. Information processing moreover offers a multitude of possibilities for the enrichment of classical products (e.g., customer service, operator guidance, teleservice) and the creation of new information markets (e.g., database services, media, World Wide Web). In addition to the increase in productivity through the facilitation of coordination and specialization we realize as well the increase in customer efficiency through quality improvements and new services.

## 1.3    Business Information Systems in the Context of Different Subjects

*Business information systems* is understood as an interdisciplinary subject between business administration and computer sciences and comprises also instructional and research matters in the field of technology. It offers more than merely the intersection among these disciplines (cf., fig. 1.3/1) such as methods for the coordination of enterprise strategies and information systems (see section 7.1).

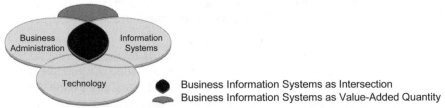

*Fig. 1.3/1      Classification of Business Information Systems*

To a large extent we agree with the Association for Information Systems (AIS), the leading international association of information systems: While economics, business administration and related social sciences tend mainly to focus on the important, classical production factors of capital and labor/work, business information systems concentrates on the production factor of information, i.e. the central resource of the information society.

Information processing penetrates all sectors of economic life. This is also reflected in that business information systems as a subject has tight relations with many disciplines (cf., fig. 1.3/2). When considering the business subfields of commerce, trade, banking or insurance as instructional subjects, then this corresponds with the (integrated) information processing in companies within the business of commerce, trade, banking or insurance.

The structuring of *functional areas* and *processes* that is so common within the field of business administration finds its equivalent in business information systems: Here one developed, e.g., AS for computer-supported research and development, for information processing in sales, for largely automated manufacturing and also for Computer Integrated Manufacturing (CIM) (see section 5.1.5.1), as well as partially automated customer information systems such as help desks. Also modern controlling, a function cutting across many departments and areas, as well as external accounting systems and human resources are today inconceivable without the use of computers.

Teaching *management decision-making* has many ties to business information systems. Management support systems (see section 4.3.2.2) show at which points decisions are required and help to prepare them, e.g., through the simulation of alternative strategies.

Some modern forms of the entrepreneurial organization, especially the network organization or the virtual enterprise (see section 7.1.1.1) can only be realized through information systems in the first place.

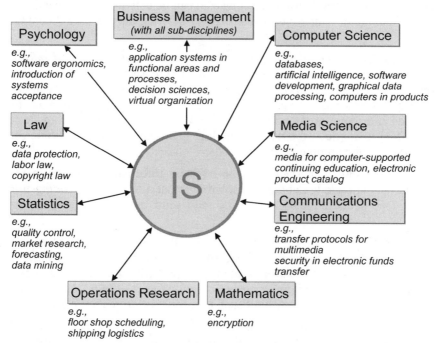

*Fig. 1.3/2     Business Information Systems in the Subject Matter Canon*

The relations of business information systems to *informatics* or *computer science* are very close. Administrative systems, as well as disposition, planning and controlling systems (see sections 5.1.12 and 5.1.13) derive their data from *databases* that have been developed by informatics specialists or computer scientists.

The most modern systems within the field of business information systems benefit tremendously from the demanding developments in informatics or computer science. Elements of the subfield of *artificial intelligence (AI)* can be found within elegant disposition systems. Many procedures advanced within the field of *software development* are the basis of application systems development in business information systems. As another example we refer to *graphic data-processing* which permits the graphical depiction of products, e.g., furniture, on the computer display unit. When deciding upon a product strategy it behooves to consider computers that automate the use, e.g., of a video camera through an embedded system.

More recent technical progress enabled the integrated use of previously separate and isolated *media* (numbers, texts, graphics, still and moving pictures, sounds) within the same presentation. Business information systems utilize this multimedia technology, e.g., when user handbooks for production machines are no longer printed, but instead can be activated on demand in the form of sound or video-based instructions directly on the user's computer screen.

*Communication engineering* is available for the *transmission of information* in local, regional and global networks (see section 2.4). In order to transmit *multimedia information* in solid quality and securely between computers and user terminals business information systems continue to place new demands and requests upon the field of communications engineering. A corresponding example is the use of a mobile phone during the electronic sales process ("mobile commerce").

From the field of *mathematics* the business information systems person, e.g., derives knowledge about the possibilities and limits of *encoding* information (see section 2.5.4) that firms may send over wide area networks (WAN).

The field of *operations research*, i.e. the discipline focusing on assisting decision-making by mathematical algorithms, one may view as a sister discipline of business information systems. Improved methods of computer-assisted *shop floor scheduling* or *transportation logistics* (e.g., the disposition of trucks) cannot function without such procedures (see section 4.3.2.2).

Similar things may be said about the field of *statistics*. It is needed for computer-supported *quality control, market research,* the *forecasting* of orders and shipments or for the searching through very large data bases such as in data mining (see section 4.3.2.2).

A good business information systems specialist has to be familiar with some areas of *law*. Here we are thinking, e.g., about such areas as *data and privacy protection, employment-related laws*, as well as *copyright laws* when developing new information systems.

Maybe a novice to this field may think that a sober and formal discipline such as business information systems has little to do with *psychology*. This, however, would be a mistake. Psychological studies are needed, e.g., during the development of new computer display surfaces (*software ergonomics*, see section 6.2.1.4). The question pertaining to under which conditions the *introduction of an information system* (that alters severely how work is conducted) can be accomplished and when it fails is the subject of many studies in which also behavioral scientists participate. Moreover, there is always the challenging question asking when executives are either welcoming computer support or reject such support outright (*acceptance problem*).

## 1.4 Structure of the Book

In order to evaluate, develop and maintain application systems, as well as to be able to build such systems into integrated information systems, one not only needs to have a solid knowledge of business administration but also an array of additional knowledge:

- The technical tools are essentially the computers that we encounter in our daily life typically as the personal computer. Beyond that they are used in order to provide the user with network services and connections. In chapter 2 the reader will be offered a first introduction to the material and immaterial components of systems. In the subsequent sections we will explore first the important characteristics of various *computer classes*, followed by an overview of *network concepts* and *network architectures*. The final section describes the *Internet*.

- In chapter 3 we address the depiction, storage, evaluation and integration of *data*. First we explicate the structure of data and databases and, building upon this, we then address the linkage of different data bases in networked systems.

- Application systems may support different functions and processes which need to be integrated and which process common data. In chapter 4 we demonstrate the methods of integration and create a comprehensive toolset to map application architectures to integration and enterprise models, respectively. These, in turn, will be dealt with in chapter 6.

- In chapter 5 we sketch the content of important application systems and their integration. Our selection is such that on the one hand we cover the most commonly occurring business application systems and on the other hand we will describe through the use of examples the most important techniques, such as dialogues or electronic mail. Since methods are also dependent upon industry or business type, we are differentiating among enterprises that produce physical goods (manufacturing firms) from those that provide services (trade, banks, transportation and gastronomy). Subsequently we address electronic commerce, i.e. the interaction and transaction among firms, e.g., in the area of sales of products and customer relationship management, based on newer electronic media. Finally, we offer an example of inter-enterprise integration.

■   In chapter 6 we learn through which phases and with which tools one may *plan for* and *realize* an application system within a project management framework. All along it is important to focus in this process on the role of the human being as the developer or user of an application system.

■   Information systems influence today the success of an enterprise in fundamental and decisive ways. This is why the information systems strategy and the entrepreneurial strategy together with the firm's structure have to be aligned well. The business information systems area within the firm needs to be organized efficiently. These tasks are referred to as information management (chapter 7).

## 1.5   Literature for Chapter 1

Mertens 95          Mertens, P., Wirtschaftsinformatik: Von den Moden zum Trend, in: König, W. (ed.), Wirtschaftsinformatik '95, Wettbewerbsfähigkeit, Innovation, Wirtschaftlichkeit, Heidelberg, Germany 1995, pp. 25-64.

Picot et al. 01     Picot, A., Reichwald, R., Wigand, R., Die grenzenlose Unternehmung, 4th edition, Wiesbaden, Germany, 2001.

Wallis/North 86    Wallis, J.J., North, D.C., Measuring the Transaction Sector in the America Economy, 1870-1970, in: Engerman, S.L., Gallman, R.E. (eds.), Long-Term Factors in American Economic Growth, Chicago/London, 1986.

Wigand et al. 97   Wigand, R., Picot, A. Reichwald, R., Information, Organization and Management: Expanding Markets and Corproate Boundaries. Chichester, England: Wiley, 1997.

# 2 Computers and Networks

Computers serve the automation of information processing efforts and the support of decision-makers. They get integrated into networks in order to increase the overall utility. New information and communication networks—at the forefront the public Internet as the "network of networks"—lead today to profound changes that influence working life (e.g., the outsourcing of entire production steps abroad), as well as all segments of social life (e.g., online shopping).

In this chapter we describe the technical foundations of these networks. After a depiction of material and immaterial components of microcomputers, i.e. hardware and software of personal computers (PC), we explain important characteristics of different classes of computers. An overview follows addressing network concepts and architectures that integrate computers. The conclusion offers a description of the Internet, the infrastructure of global networks, which make worldwide information resources accessible cost-effectivly and cooperative processes executable.

## 2.1 Hardware

Under the designation *hardware* we simply understand all equipment that the user can "touch", i.e. which possesses material characteristics.

A typical PC workplace consists of the following hardware components:

- The central processing unit (CPU) being comprised of the processor and the main memory devices
- External memory or storage devices (e.g., hard-disk, diskette, micro-disk)
- Data entry devices (e.g., keyboard, mouse, scanner)
- Data display devices (e.g., computer display screen, printer)

Such a workplace is regularly enhanced by being connected to a network (e.g., via network card, modem) (see section 2.4).

Aside from these stationary work places portable PCs are enjoying ever-increasing usage. Most users have *notebooks*, roughly the size of an 8 ½ x 11 inch page. Pocket computers are yet smaller; they are referred to as *Personal Digital Assistants* (PDA). Moreover, increasingly *mobile phones* offer functions and features resembling those of pocket computers.

The way a computer works is as follows: Data need to be *input*, e.g., via the keyboard, optical reader or external memory device. These are then *processed* yielding results in the form of *output* on the computer display unit, the printer or on an external memory device. Working this way is referred to as the Input-Processing-Output (IPO)-Principle.

A *central processing unit* (CPU) consists in its basic configuration of *a* single main memory unit and *a* single processor which by itself is comprised of *one* compute unit and *one* control unit. Sometimes the term CPU is used also to refer to the processor by itself. Figure 2.1/1 clarifies this architecture.

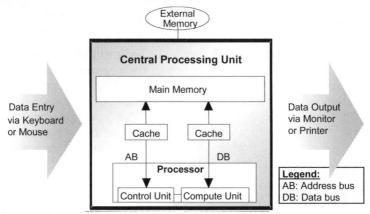

Fig. 2.1/1          *Generalized Structure of a Central Processing Unit*

A task submitted by the user to the CPU is fulfilled as a sequence of state-transitions of the main memory unit. This idea is demonstrated through the example "Movement through the maze" in figure 2.1/2 in which the processor can only interpret and execute moving commands from the programming language SUPERSTEP. A command describes the moving direction (north, west, south, east) as well as the number of steps (1 step, 2 steps) according to coding instructions. A command is expressed in a combination of bits, with one bit being a single symbol expressed as either a 0 or a 1. A command is comprised of <Bit 1, Bit 2, Bit 3> with Bit 1 and Bit 2 specifying the direction of the movement and Bit 3 determining the length of each step. Thus a program can be formulated that accomplishes the task depicted in figure 2.1/2. The sequence of commands is entered into the left column of the table and the resulting new location of the person after each execution of the command is entered into the right column.

As an elaboration of the program's functioning as a sequence of main memory state-transmissions, we imagine that the entire program is loaded into the main memory and the starting location of the person is initialized. The *control unit* reads the first command from the memory. Then it interprets the command and instructs the *compute unit* to carry it out. For this purpose the compute unit reads the location from the main memory unit and alters it as instructed. The result of the execution of the first command is being deposited again in the main memory unit (e.g., by writing over the "old" position with the "new one") and the control unit takes the next command in order to proceed as described above. It interprets the mode of operating, fetches the operand (here the location of the person) from the main memory unit and carries out the operation by changing the content of the main memory unit. Computers whose CPU is configured like figure 2.1/1, carrying out the just described processing steps for a sequence of commands, are named after the mathematician and cybernetician John von Neumann and one refers to them as *von Neumann computers*. These principles of the so-called memory-programmable computers were developed in the middle 1940ies. A depiction of alternative computing architectures that, e.g., may offer several parallel operating processors can be found with [Giloi 93], among others.

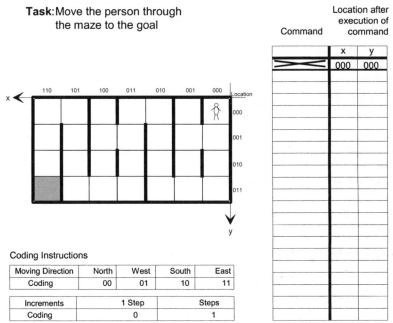

Fig. 2.1/2     *A Maze and Coding Instructions*

Since the hardware delivers "merely" the general processing mechanism for the command structure it is considered multi-functional. Additional systems software, but also data (see chapter 3) are required for any targeted processing operation.

In the following sections we explain the central processing unit, external storage, data paths as well as input and output devices.

## 2.1.1  Central Processing Unit

### 2.1.1.1   Processor

Manufacturers of microprocessors express the performance of their devices through the measurement unit megahertz (MHz) which specifies the cycle time frequency of the processor. It determines how many commands per second will be executed, but it does not give us an easy indication about its processing speed. This is influenced by other factors, including the processor architecture and the mix of programs that have to be run.

The internal processing speed of a processor depends on how quickly, e.g., the individual components—compute unit, control unit, cache and main memory—operate and moreover how quickly communication takes place between these components. They are connected via so-called *busses*, consisting of multiple cable veins. One differentiates between the address bus that connects the main memory with the control unit (cf., fig. 2.1/1) and the data bus that connects the main memory with the compute unit. The bit statement (an additional performance measure of a processor) generally refers to the number of data that may be transmitted simultaneously via a bus (see section 2.1.3). With wider busses one attempts to achieve higher processing speeds. For example, Intel processors of the type Pentium III have a 64-bit data and a 36-bit address bus.

Since the processing speed of a processor usually is higher than the access speed of the main memory unit, caches (which may be components of both the processors and the main memory) hold those kinds of information that the compute unit is likely to require in the next steps of processing.

### 2.1.1.2   Main Memory

The *main memory* of a data processing system consists of the random access memory (RAM) and the read only memory (ROM).

The *random access memory* is comprised of directly addressable memory cells called memory words. In a PC a word is usually 2 or 4 bytes (1 byte has 8 bits), with mainframe computers this is usually 4 bytes. Main memory capacities are specified in megabytes (1 MB = $2^{20}$ byte = ca. 1 million bytes). Today PCs have generally a capacity of 128 to 1024 MB (=1 GB).

All programs have to be fully or partially (namely the actual portion to be executed) resident in the main memory at the time of their execution. In the latter case the operating system offers the *virtual memory technology*. Program portions that no longer can be kept in the working memory (as other programs, e.g., also need to be loaded for quick execution in the main memory) are stored automatically on the hard drive. They will be imported only on demand into the working memory, which thereby grows logically, but not in its physical size. This im- and exporting of information on the hard drive is also referred to as *paging*, since programs and main memory consist of several equally large pages. In contrast to magnetic storage devices that keep storing data even after the computer has been turned off, the main memory loses all stored information when the electricity is turned off.

A read-only memory (*ROM*) permits merely reading of information, i.e. it is impossible to change any of the stored data. ROM is usually created by the PC's manufacturer. An alternative procedure is to have users write once into a ROM (programmable ROM = PROM). Finally, there is a ROM whose content can be erased and may be replaced through new information (Erasable PROM = EPROM). Read-only-memory serves among other things the storing of fundamental portions of the operating system which are read automatically (e.g., hardware-dependent programs for the controlling of the display screen or for communication with the keyboard) when the computer is turned on initially.

## 2.1.2    External Memory

*External memory* is especially suited to store larger data amounts on a long-term basis and to make it transportable. Important external storage media are:

- Hard-disk
- Diskette
- Magnetic tape or streamer
- Optical storage devices
- Smartcard
- Microdrives
- Thumbdrives

A *hard-disk* is usually comprised of several layers of plastic or aluminum disks that are covered with a layer that can be magnetized. Often, one also refers to such a system as stack of disks. Data are stored magnetically in the form of bit chains in concentric tracks. The usually permanently installed stack of disks turns at a consistent speed. Data are read or stored by read and write heads that are being positioned radially onto the desired track. There they wait for the sector to pass by that contains the data to be retrieved.

Today magnetic hard disks for PCs generally offer a storage capacity ranging from 10 to 120 *gigabyte* (1 GB = $2^{30}$ Byte = ca. 1,000 MB). In mainframe systems (see section 2.3.1) capacities of several *terabytes* (1 TB = $2^{40}$ bytes = ca. 1 million MB) are possible. Magnetic disks offer the following advantages:

- High storage capacity
- Relatively quick access

■ Reusability, since data can be written over

■ Relatively high data security

If, e.g., copies of data sets are required as backup or if such data sets are to be exchanged between computers without using networks (see section 2.4), one usually would choose transportable storage media, magnetic tape, as well as various optical storage media.

The *diskette* is a widely used storage medium for PCs. It is a 3.5 inch in diameter disk in a hard plastic cover and a storage capacity of 1.44 MB that functions quite similarly as the earlier described hard-disk. Augmented diskette media differ in their respective speeds, formats and capacities ranging from 120 MB up to 2 GB with low access speeds (e.g., LS, Zip or Jaz diskettes).

*Magnetic tape* consists of plastic material covered by a magnetizable layer on which data may be stored. For PCs magnetic tape exists in the format of cassettes that also may be referred to as *streamers*. The cassettes have a very high storage capacity of up to several GB. Magnetic tape and cassette, however, have the disadvantage of permitting only a sequential access to the stored data. In order to read a particular data set, it is necessary to read over all the stored data one after another which leads to a long waiting period. Due to this, magnetic tape is mainly used for data security purposes, e.g., for data backup in preparation to restore a data base in case of an unanticipated destruction of the data (see section 7.3.1.1).

*Optical* storage utilizes the technology of reading the stored data under the transparent protective layer of a CD (Compact Disk) via a laser beam. Data are stored and read there via such a laser beam. Since laser light has a very short wavelength and can be positioned very precisely, an optical storage device posses a very high capacity. We differentiate among the following technologies:

■ *CD-ROMs* (Compact Disk Read Only Memory) have a storage capacity of up to 700 MB (about 80 audio minutes). Standard software packages are often distributed via this medium. Another application area using this technology is the administration of large and usually not changing data sets, such as patents, records and books. In doing so, the integration of text, video and audio becomes increasingly attractive and important. Additional developments of this technology include CD-R (CD Recordable) for once-only writing but also CD-RW (CD ReWritable) permitting the repeated re-writing onto the same CD. The storage capacity of a CD-RW is 650 MB.

■ *MO*-(Magneto Optical) Disks differentiate themselves similarly in being recordable just once (MO WriteOnce) and being re-recordable or re-writable (MO ReWritable) media. The storage capacity reaches up to 5.2 GB.

- *WORM* denotes Write Once Read Many. Via a combined magneto-optical process these media may be written once by the user and be read many times. The capacity of such a storage medium ranges from 600 MB up to 3 GB. In comparison to the CD technology its usage is very small, which may be explainable by a missing dominating standard.

- *DVD* (Digital Versatile Disk or Digital Video Disk) is the most recent development within the area of optical storage. The DVD was designed for complex audiovisual applications and serves as a major alternative for interchangeable and tape-based storage technologies. Since DVDs may carry data on both sides, storage capacities of 4.7 GB and up to 17 GB can be achieved. Further developments include the DVD-R (DVD Recordable), as well as DVD-RAM and DVD-RW for repeated writing.

*Smartcards* (see sections 2.5.4, 5.2.7.2 and 5.2.8.2) have the format of a credit card and usually carry an embedded writeable microchip. In order to provide the chip with data, special loading devices (in case of information retrieval reading devices) are utilized. The capacity of smartcards at the present time (in terms of text) is about 15 typewritten pages. Principally, we may identify three different smartcard variations:

- *Memory cards*, such as the CompactFlash (CF) card, store data and may be used, e.g., as a health insurance ID card

- *Processor cards* use their own micro-processor including an operating system (see section 2.2.1.1) and store and process data (e.g., phone chip cards)

- *Encryption cards* have an additional processor, a so-called co-processor, for the encryption of data that may be used, e.g., in conjunction with payment systems

The *microdrive* has the same size as a CompactFlash (CF) memory card, designed specifically to fit in a CF slot. It weighs only 16 grams (about half an ounce) yet can hold around 250 MP3 tunes at the standard 128 kbit compression. IBM's top of the line microdrive weighs less than a roll of 35 mm film and has storage capacity of one GB. Microdrives are typically used in conjunction with handheld PCs, laptop computers or digital cameras.

A *thumbdrive* is a USB removable pocket hard drive. It is available in 8MB-512 MB and combines flash memory technologies with the ubiquitous USB connection to create a self-contained drive and media package in the size of a thumb. It plugs directly into the USB port of any computer and can store virtually any digital data from documents and presentations to music and photos.

## 2.1.3    Data Paths

One distinguishes between internal and external data paths. The *internal* data path serves the transport of information within the CPU, e.g., between processor and the main memory. The *external* data path enables the transmission of data between the peripheral devices, e.g., external memory devices and the working memory. For internal and external data paths in micro-computers mainly busses are used (see section 2.1.1.1).

Mainframes use primarily the so-called *channel concept* for internal and external communication. One may envision a channel as a processor, specialized on data transport that runs parallel to the central processor. Also, the channel requires a program (the channel program) that is resident in the main memory, i.e. central processor and channel processor(s) work within a common main memory.

## 2.1.4    Data Input and Output Devices

The most important device for data entry aside from the *keyboard* is the *mouse*. The mouse is about fist-size and movable on the work desk area. The pointer on the screen changes as the mouse is being moved. The sensors of this movement might be a mechanical sensor such as a rolling ball or an optical sensor. The mouse usually possesses several keys with different functionalities to trigger actions. The track ball functions similarly and may be envisioned like a mechanical mouse laying on its back.

In computer games often the *joystick* is being utilized. Via a lever the pointer may be moved. Trackballs and joysticks both have keys that can trigger actions when depressed.

The *lightpen* is a pen-like input device that utilizes a light-recognizing sensor in order to choose objects on the screen.

In order to keep the usage of a computer rather simple and fitting natural human behavior the *touch screen* is increasingly used for data entry. The user points on an object and optical or mechanical sensors register the touch, as well as positioning on the screen, e.g., with an automatic teller machine (ATM). Moreover, devices exist (e.g., in the form of whiteboards) on which various digital information can be handled. Input and output may occur through boards operating with background-projection of objects on which one may write with one's fingers directly.

Another important input device is the *optical reader*. It captures normed data such as bar codes or OCR (*Optical Character Recognition*) Code. This may occur, e.g., through the use of a pen reader 'reading' such information, meaning that light vs. dark variations are recognized and evaluated. Such devices are used by scanner registers in supermarkets or in financial institutions in order to read forms. A variant of optical readers are so-called *scanners* that dissect the presented information (picture elements) within photos and graphics for example in pixels (black and white, but also in color).

*Cameras* and *microphones* are becoming increasingly popular for data entry and can be found with multimedia applications.

*Mobile data capturing devices* may be seen as smaller input and output devices, e.g., within the cabin of a truck in order to provide receipts for accepted shipments.

The most important output device for business information systems is the *monitor*. It serves data output (e.g., in the form of tables and graphics) and also supports data entry, as on the monitor forms for data capture and icons for the activation of programs are depicted. Devices in different sizes exist, with varying capabilities to use colors and varying levels of resolution. The resolution provides the number of different pixels through which graphics and pictures may be represented on the screen. Common screen sizes today are 17, 19 or 21 inches measured diagonally. Each PC requires a *graphics card*. It determines—together with technical capabilities of the used monitor—the degree of resolution, as well as the number of presentable colors.

Another important output device is—aside from the monitor—the *printer* through which it is possible to represent results onto paper. Inkjet as well as laser printers are used. Inkjet printers create letters and graphics using many individual dots which are squirted at very high speed on the paper. The laser printer has a much higher print quality, as well as a higher speed. The page is being build as a whole in the printer and is being transmitted onto the page by using toner. This procedure is quite comparable to that of photocopying machines.

An additional output device is *computer output on microfilm*, enabling the capturing of documents for archiving purposes in a space-saving medium.

*Virtual reality*, an artificially created environment through hardware and software, utilizes, e.g., *data gloves* as an input device. Through the movement of their hands users have the opportunity to maneuver within a simulated space—observable with special eye glasses or a helmet—and to bring about a series of specific actions. One possible application of this technology, among many others, is the sale of house furniture, enabling the moving and placing of furniture pieces within a predetermined space.

Aside from using these devices for the initial capture of data increasing importance is found with the supply of information by another computer over a network (see section 2.4). This applies also to the data output to another PC.

## 2.2   Software

Software is the necessary prerequisite for the operation of a computer. The term generally refers to written programs within a particular programming language (see section 2.2.1.2) that are executable after being interpreted by a computer. Based on the criterion of nearness to the hardware or nearness to the user one differentiates, respectively, between *system software* on the one hand and *application software* on the other hand.

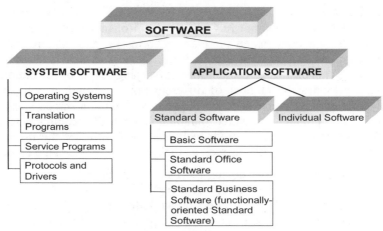

*Fig. 2.2/1      Classification of Software*

A focal demand on system software is to make the hardware more useable. For example, it would be uneconomical to build one's own printer drivers for each application program that address, e.g., the eventuality that the printer has run out of paper. Furthermore, similar programs for numerous adminis-trative and control tasks are referred to as operating systems.

*System software* comprises—aside from the operating system—translation programs (for varying programming languages), utility programs (often used programs, e.g., for the sorting of data), as well as protocols and drivers (for communication with peripheral devices and other computers within a net-work).

*Application software* may be grouped into two classes: individual *software* (e.g., for the controlling of a luggage routing system at an airport) is being specifically constructed on the wishes of a particular user (other department or firm). *Standard software* is designed having a multitude of users in mind with the same or a similar problem orientation. From a business administra-tion point of view, standard software may be subdivided into *functionally independent* software, so-called *basic software* (e.g., web browsers) and *standard office software* (e.g., word processing), as well as *functionally-oriented* software (e.g., cost accounting).

With microcomputers the use of standard software dominates whereas with mainframes (see section 2.3.1) individual software still enjoys considerable importance.

The boundaries between these software categories are somewhat fluent. For example, database systems (see section 3.1.5) with microcomputers are seen as standard application software, yet with mainframe machines they are understood as system software.

Especially in the private consumer use area further standard applications exist such as off-the-shelf *entertainment programs* or computer games, but these are not part of the following considerations.

## 2.2.1 System Software

The first section describes operating systems for microcomputers. In the following we explain and classify programming languages. Based on this discussion we continue describing interpreters and utility programs.

### 2.2.1.1 Operating Systems for Microcomputers

The *operating system* has initially the task to coordinate independent components (e.g., central processing unit, printer, keyboard) in conjunction with solving a task. Operating systems are the interface between the user, respectively the application software, on the one hand and the hardware on the other hand. They have to meet the following requirements:

- Provision of file management (see also chapter 3)
- Administration of hardware resources (processor, main memory, peripheral hardware)
- Administration of user tasks and control of program executions
- Provision of (graphical and text-based) interfaces enabling the user to communicate with the system

Modern operating systems for microcomputers may be characterized by the following features:

- Graphical user interface (GUI)
- Hierarchical file management
- Possibility for batch and transaction processing
- Multitasking

Beyond this description some operating systems support so-called *multiusing*.

The term *hierarchical file management* is well known from classical paper organization in offices. For example, a filing cabinet consists of several shelves of filing drawers (e.g., for invoices, receipts, etc.). On each shelve one finds individual folders with stored records in an orderly fashion.

The data processing system collects and stores not only data but also methods (programs), specifying how these are to be processed. In doing so files (here defined as collection of data logically belonging together, see section 3.1.4) are being administered by an operating system on external storage media (e.g., on a microcomputer's hard disks or diskettes). A user may create there directories and sub-directories to speed-up access.

In figure 2.2.1.1/1 the user has created, e.g., on a hard-disk a main directory of all directories (here Internet, texts, graphics, data base and system). The folder labeled "data base" is subdivided into the sub-directories ACCESS™ and ORACLE™. One may envision the formal structure of a tree with its root (on the left side) and varying leaves (on the right). Programs and data are attributed to and are available in the leaves, i.e. in the respective directories.

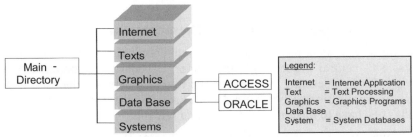

*Fig. 2.2.1.1/1    Hierarchical Arrangement of Directories*

One speaks of *batch processing* when a sequence of commands (embedded in a file) is being processed one after the other without any influence by the user. This sequence of commands needs to be fully specified prior to the start of the batch processing. An example is the printing of a payroll.

The counterpart to batch processing is *transaction processing*. Here the user merely submits partial tasks and stays in constant contact with the central processing unit through some dialog (e.g., if the system expects the submission of a command, this has to be given and will be executed immediately). Such a form of processing may also be referred to as interactive. An example of this is the entry of a client order in an input mask.

If the operating system supports so-called multitasking the computer is capable of executing several programs at the same time. For example, it is possible to work on a text, and while the computer is waiting for the next command the PC is processing a calculation in the background. In this context one also speaks of *multi-threading* when an operating system permits a program consisting of several processes, i.e. threads (e.g., a print process and a computation process), and handles these processes in a quasi-parallel fashion.

*Multi-using* exists when a central processor serves several terminals and thus several users in a quasi-parallel fashion. In contrast, one speaks of *single-using* when only one user is being served.

Personal computers today use mainly operating systems by Microsoft™ (MS) which have emerged as an unofficial standard. *Windows XP, NT or 2000* permit multitasking and offer a joint utilization of resources in networks (see section 2.4).

*UNIX systems* allow for multitasking and multi-using. Moreover, some UNIX variants provide an integrated software development environment. The term UNIX suggests a uniformity that is not reflected in the marketplace. Several variants of UNIX and manufacturer-specific implementations (e.g., Solaris from Sun Microsystems™, AIX from IBM™, HP-UX from Hewlett-Packard™) exist. Some of these operating systems support multi-threading.

A particular case among these Unix-derivatives are *Linux* operating systems which are—in contrast to commercial systems—freely downloadable by anyone (e.g., from http://www.redhat.com/download/). These systems continue to be expanded and improved in a so-called *Open Source Community*, a community of experts and supporters. They are fully available in the form of the original source program. This offers specialists the opportunity to make their own improvements and modifications, to test the system, e.g., for unwanted security lags and to participate in the further development of the system.

In the area of handheld computers a number of other operating systems have been developed, characterized by their high performance and low energy consumption. Among the most commonly used systems are MS Pocket PC, PalmOS or Embedded Linux.

## 2.2.1.2 Programming Languages

A computer and its operating system support the user to solve a variety of different tasks (e.g., accounting, planning). Based on the operating system interface application systems need to be constructed that provide such solutions. *Programming languages* are used to create application systems (as well as the operating system itself). A programming language is a formal language for the development of software that runs on specific hardware.

In the past, one often classified programming languages according to so-called generations, but today such a classification occurs more often through *programming paradigms*. The unambiguous attribution to a single paradigm is difficult since individual programming languages have characteristics of varying approaches. In the following we address the characteristics and attributes used to classify programming languages.

In an imperative language the programmer defines step-by-step how a task or problem is to be solved. Programs consist of a set of commands and control structures that define wether statements are executed sequentially or in parallel, resulting in state-transitions. The state base is built on constant and variable values.

Subsets of imperative languages are *procedural* and *object-oriented* programming languages. Procedural languages are oriented on the von-Neumann computer architecture and use separately designed data and command structures. Such programming requires appropriate knowledge and experience. It is possible, e.g., to align the necessary coding in a domainspecific language to solve a certain problem and to perform it in a largely machine-independent programming language. One refers here to a problem-oriented programming language. An important advantage is that users can understand many of the program constructs. Moreover such programs are transferable (maybe with small modifications) to other operating system environments. Widely used problem-oriented and procedural programming languages are, e.g., BASIC, C, COBOL and FORTRAN.

Computer hardware, however, is not immediately capable to "understand" commands from a programming language (source code). First they have to be transfered by using a translation program (see section 2.2.1.3) into *machine language*, i.e. into binary code with the delineation of commands and data (object code) which can be readily processed by the hardware (cf., figure 2.1/2).

In contrast, *object-oriented (OO) programming languages* describe a program as a collection of objects that are interconnected and capable to exchange information for the purpose of problem solving. Objects of a similar type belong to the same class. For each class it is predetermined which conditions objects may accept and which changes of the object condition may be executed when information arrives. The changes are made by methods that are tied directly to the object (see section 6.4.1.4). OO languages support the abstraction process by classification and encapsulation of object data with those procedures which are allowed to alter the object's data. Together with the concept of inheritance software components can be reused and thus contribute to an increase in productivity within the field of software development.

An example of a business object is an invoice that has a predefined data structure (e.g., the invoice header with the recipient of the merchandise and the invoice line items with the item identification number, quantity delivered and price). The object "invoice line item" defines again the permissible methods, e.g., to generate a new item in the first place or to change one.

Over time, two different approaches have been pursued in the provision of OO concepts in the area of programming languages. One concept contains the development of new programming languages based on the object-oriented paradigms; they are being referred to as object-oriented languages. Examples are SMALLTALK and Java. The latter language constitutes insofar particularity as varying characteristics of object-orientation have been diminished, e.g., the concept of multiple inheritance. Alternatively, the enhancement of traditional programming languages through OO concepts is observable. Representatives of these so-called hybrid languages are, e.g., C++ (the object-oriented expansion of C) and VISUAL BASIC (containing OO concepts since version 4.0).

The object-oriented programming language Java increasingly gains importance for network-based application systems. In addition, it offers the possibility to devise programs that can be executed within a web browser (applets, see sections 2.2.1.3 and 2.5.2).

Aside from the imperative programming languages, *declarative* programming languages can be distinguished. One important characteristic is that the user no longer needs to formulate HOW a particular problem is to be solved, but merely needs to specify WHAT has to be solved. The translation program (see section 2.2.1.3) then adds the procedure, i.e. the concrete operational sequence for the solving of the WHAT task.

Subsets of declarative languages are functional and logical programming languages. The functional paradigm is based on the description of programs like mathematical functions. They consist of a set of functions composed by specific rules. An example is the programming language *LISP (List Processing Language)*.

Within the logical paradigm programming is viewed as providing proof of facts and reasoning. Especially the so-called first order logic [Bremer 96] is being applied here. One feature often mentioned in literature is that application systems are described in the form of rules (see also section 4.3.2.3). One rule is the causal relationship between a (complex) condition and a conclusion (e.g., *if* the buyer is known and his/her credit standing is certain, *then* deliver goods on account!). *PROLOG (Programming Logic)* is one of the best known logical programming languages. Logic and functional languages are often also referred to as knowledge-based languages that have gained importance in conjunction with artificial intelligence (AI) (see section 4.3.2.3).

Moreover, query languages for database systems (see section 3.1.9) also belong to the category of declarative languages. They are easily learnable and are user-friendly, even for users with little programming know-how. Examples of such programming languages are SQL (Structured Query Language), the quasi-standard for relational data bases (see section 3.1.9), and ACCESS BASIC for the data base system MS Access.

Figure 2.2.1.2/1 captures the various paradigms at one glance.

| Paradigm / Language | Impera-tive | Proce-dural | Object-oriented | Declara-tive | Func-tion-oriented | Logic-based | Rule-based | Knowl-edge-based |
|---|---|---|---|---|---|---|---|---|
| C | ✓ | ✓ | | | | | | |
| C++ | ✓ | ✓ | ✓ | | | | | |
| Lisp | | | | ✓ | ✓ | | | ✓ |
| Visual Basic | ✓ | ✓ | ✓ | | | | | |
| Pascal | ✓ | ✓ | | | | | | |
| Prolog | | | | ✓ | | ✓ | ✓ | ✓ |
| Java | ✓ | | ✓ | | | | | |
| Smalltalk | ✓ | | ✓ | | | | | |
| SQL | | | | ✓ | | | | |

*Fig. 2.2.1.2/1   Programming Languages and Paradigms*

## 2.2.1.3    Translation Programs

The translation of a source program to an object program occurs through a compiler or an interpreter.

*Compilers* translate the entire source program "in one piece" (batch). They examine first the program at hand for syntax errors, e.g., the correct spelling of all commands. In the next step the program will be compiled. It is then, however, not directly executable and has to be expanded through the *linker* via support programs (e.g., input and output control) that are stored in libraries. An advantage of compiling is that, based on the broad overview of the complete source program, an optimization of the object code may occur and that the executable program can be stored (e.g., on a hard disk) such that it may be loaded without delay directly into the main memory for execution. In addition, a separate compilation of enclosed modules of an entire program package is possible. This simplifies the testing of those modules. The separately compiled modules may be linked subsequently to an executable program. A disadvantage of compilers is that modules must be newly recompiled after an error correction or program change occurs.

*Interpreters* in contrast do not generate archivable object code. Each command is processed individually, i.e. the conversion occurs each time anew and the command is being executed immediately. This procedure offers advantages for interactive program development. In this way one may examine the program's correctness step-by-step.

In some programming languages the described procedures are being combined. For example, the conversion of a Java source code occurs in two steps: A *compiler* transforms the source code into an intermediate code (byte code) which was conceived specifically for an efficient and safe transmission within the network and that may be executed on any hardware and system software platform for which a *Java Virtual Machine* (*JVM*) exists. The JVM may be integrated in a web browser. In this way it is possible to integrate program building blocks that may be distributed worldwide to one executable program.

### 2.2.1.4     Utility Programs, Protocols and Drivers

*Utility programs* are support programs for the processing of system-oriented, frequently reoccurring, application-neutral tasks. These include in particular:

- Editors

- Sorting programs

- Other support programs

An *editor* is a program used for reading, changing and writing of files, e.g., with formatted data, text and graphics. *Sorting programs* serve the sorting of data based on user-specified criteria.

Additional *support programs* fulfill functions such as user-friendly copying of files, data protection, the optimizing of storage organization, etc. Support programs may be found as a component embedded within the operating system (e.g., programs for defragmenting the hard disk in MS Windows), as well as basic software (e.g., Norton™ Utilities).

Also, implementations of protocols and drivers (see section 2.4.1) may be interpreted as utility programs. Protocols define all agreements and procedures required for the communication between processes and computers. A driver is a program that acts as an interpreter between the protocols of varying functional units or a program and a functional unit (e.g., a printer). Data submitted by a functional unit are being transformed or adapted into the internal data representation schema of a computer and vice versa, signals from a computer are translated into the format of a functional unit. Protocols and drivers are often delivered with the operating system (see section 2.2.1.1).

## 2.2.2     Application Software

According to figure 2.2/1 this section deals with the various facets of standard software, followed by the characteristics of individual software.

### 2.2.2.1     Standard Software

Standard software as a term has been labeled in various ways. The reader may encounter among others: (commercial-)off-the-shelf software (or (C)OTS), packaged software, shrink wrap (as well as click wrap) software, and the clunky but descriptive label software-for-a-market.

First, we focus on basic software that offers fundamental services for the utilization of a PC workplace (also when connected to a network), within the enterprise as well as in a user's home. Subsequently we describe standard office software that is independent of specific tasks (e.g., word processing). In the last section we will discuss standard business software which supports functions of enterprises (see section 2.2).

It is common to distribute basic, as well as standard office software with programs that install themselves. Typically the user needs to specify the location of the object code within the data hierarchy of the operating system (see section 2.2.1.1).

*Functionally-oriented standard software* usually offers a multitude of support for functions, business processes, as well as business decisions. Simple software in this category (e.g., MS Quicken) provides self-installation routines without any individual customization possibilities. In contrast to the latter example parameterized standard software such as SAP R/3™ has to be customized to solve the specific needs of an enterprise.

### 2.2.2.1.1    Basic Software

A vast variety of basic software is available especially for the various MS Windows operating systems. More recently, Linux operating systems also began to provide various standard software applications, like, e.g.:

- E-mail
- Browser
- Website editors
- Virus scanners
- Compression programs

The electronic exchange of messages among people is referred to as electronic mail (e-mail). Aside from the basic functions of creating, sending and receiving an e-mail note, such systems offer features to define distribution lists, to manage contact addresses in data bases, as well as to encrypt messages. The *Simple Mail Transfer Protocol* (SMTP) controls the transfer of electronic mail between clients and servers (see section 2.4.4). Well known mail-clients are MS Outlook Express and Netscape™ Messenger.

The term *Browser* defines in general support programs which allow data recall facilities and the placement of data within a directory hierarchy. The visualization occurs typically via tree-like structures. If a browser is being utilized for an audio-visual depiction of HTML pages (see section 2.5.2) within the World Wide Web (WWW) (see section 2.5.2) one refers to a web browser (e.g., MS Internet Explorer and Netscape Navigator). An HTML document (see section 2.5.2) may include references to other documents, but also with multimedia objects such as video, audio, graphics and language. The user access takes place by the specification of the Uniform Resource Locator (URL). Web browsers serve in particular as navigational aids while searching the Internet. They may work with the use of so-called plug-ins and various help applications to explore additional data formats.

Editors for websites such as MS FrontPage, HomeSite or Dreamweaver are programs for the preparation of texts or graphics that are to be posted on the Web.

Virus scanners enable the examination of storage areas of computers for known viruses and to quarantine or eliminate these.

Compression programs serve to reduce a file to a minimum size by, e.g., substituting a sequence of *n* zeroes to be carried forward through the expression $O(n)$. The reconstituting of the original character sequence occurs through an inverse function. Compressing is especially important during the transfer of data, e.g., with e-mail, to minimize the transmission time between sender and receiver.

### 2.2.2.1.2    *Standard Office Software*

*Standard office software* based on Windows operating systems contributes to the continued and wide usage of microcomputers. This type of software includes:

- Word processing
- Presentation graphics
- Spreadsheet analysis
- Database administration (see chapter 3)

*Word processing programs* are designed to simplify the writing of texts, e.g., letters, documents, theses, etc. Beyond that nearly all word processing programs offer additional desirable functions, such as:

- Automatic pagination
- Footnote administration
- Syllabication
- Spell checker
- Thesaurus
- Serial letters

Most word processing programs permit also the integration of graphics that may have been produced with other appropriate programs. With add-on functions and features we encounter an increasing number of commands. The usage of such commands is greatly simplified in modern software packages by offering pull-down menus. Such programs can be found in nearly every office today. Well known products are MS Word and WordPerfect.

*Graphics programs* offer a multitude of illustration facilities for the conversion of cipher-oriented data into graphics. For example, the monthly sales and costs of a company can be depicted by using bar diagrams. A well known product is Harvard Graphics. Another often used type of graphic depiction is the structured grid or surface depiction of bodies in technical applications (e.g., Computer Aided Design (CAD), see section 5.1.1.1).

Programs that enable the integration of graphics with texts are referred to as *desktop publishing programs*. These programs handle presentation graphics, graphics from CAD programs or pictures that are read by scanners and depict texts in varying forms and type faces at the same time. Most publications such as newspapers, technical handbooks, etc. are being produced with the assistance of desktop publishing programs. An example is Adobe™ PageMaker.

A spreadsheet program depicts information in the form of tables, i.e. using columns and rows. The size of such a table or spreadsheet is usually 256 columns and 16,384 rows. Rows are numbered consecutively and columns are labeled through ordered letter combinations. The intersection of columns and rows are referred to as cells. These are specified—just like the fields on a chess game board—by corresponding column-row combinations. Typical applications using spreadsheets are routine calculations of various sorts, especially the business management applications such as budget, finance and personnel planning. Almost all spreadsheet programs offer a subprogram for the graphic depiction of data within a work sheet. This module makes the creation of bar diagrams, pie charts, etc. possible (similar to the earlier referred to graphics programs). Well-known spreadsheet programs are, e.g., MS Excel and Lotus 1-2-3.

### 2.2.2.1.3    *Standard Business Software Supporting Functions of Enterprises*

We refer to *functions-oriented standard software* as solutions that support from a business management perspective different functions or function-spanning application areas (e.g., materials management, sales and distribution, finance, controlling, human resources and production) and their processes. Function-spanning, integrated standard software is comprised of modules that access a common database. An advantage from the user's perspective is that only software needs to be activated that is required for his or her problem solution. In this way, e.g., only modules for lead time scheduling and capacity balancing as part of production planning and control need to be acquired, but not shop-floor scheduling (see section 5.1.5.8). The modular design enables a stepwise introduction of new systems and thus a slower replacement of older systems.

The adaptation of such standard software for specific demands in firms takes place by the adjustment of parameters without having to alter source code. Beyond that, interfaces are being offered for individual expansion.

Function-spanning standard software may be subdivided into *industry- and line-of-business-independent* or *industry- and line-of-business-specific* standard software. Examples for *function-spanning, industry- and line-of-business-independent standard software* are primarily enterprise resource planning (ERP) systems like SAP R/3™ as well as supply chain management systems such as Advanced Planner and Optimizer (APO) by SAP™ and Rhythm from I2™.

*Function-spanning industry- and line-of-business-software-specific* is merely conceived for just one user segment (e.g., the automobile industry) and encompasses usually extensive knowledge of that industry and line-of-business. Solutions, such as *SAP™ Industry Solutions*, are sold today by nearly all larger standard software developers.

According to the previous structure we may divide standard software serving just one function into solutions that are independent from industry and line-of-business (e.g., Internet-based purchasing systems) and industry and line-of-business dependent standard software (e.g., cost accounting systems for the insurance industry). Figure 2.2.2.1.3/1 offers a few examples (see section 5.2):

| Industry | Application system |
|---|---|
| Transportation | Application systems for long-term transport planning and short-term transport control |
| Tourism | Integrated travel distribution systems for bid management, reservation management, accounting and ticketing |
| Gastronomy | System to support customer service, e.g., order acceptance, bill printing and account calculation |
| Health care | Integrated medical information systems to administer patients, document treatment, as well as cost accounting |

*Fig. 2.2.2.1.3/1 Examples of Industry- and Line-of-Business-Specific Software*

## 2.2.2.2 Individual Software

Under *individual software* we understand application software that is tailored for a specific business need within an appropriate hardware and software environment. Such software may be developed by the individual firm itself or possibly bought from a vendor (for criteria in making such a decision see section 7.2.1). Within the enterprise, development of individual software may occur by the information systems department or the "application" department. In the latter case such development is likely to occur using declarative languages (see section 2.2.1.2). The task at hand here is to master this development of application software technically and financially as a single made-to-order production (see section 6.2).

Due to the high development costs of individual software we observe today even in larger enterprises that standard software is being deployed. In contrast the deployment of individual software may be justified when standard software solutions offer an insufficient functional spectrum for specific problems (e.g., the control of a luggage routing system at an airport).

# 2.3   Computer Classification

The following *computer classes* are of relevance for the configuration of computer and network infrastructure being used in business:

- Mainframe computers

- Workstations

- Network computers and thin-clients

In addition, so-called *supercomputers* exist with a special computer architecture that are usually deployed for technical-mathematical problems (e.g., weather prediction, air space surveillance). In contrast to mainframes supercomputers make their entire performance capacity available merely to very few programs. *Computers used for process control purposes* in statistical process control are single function automata that are, e.g., built robustly for the operation in a workshop. Moreover, aside from the controlling of a machine (e.g., the machining center in the automobile construction or a reactor in the chemical industry) these computers are also capable of sending data via networks (see section 5.1.5). Frequently one also refers to so-called embedded systems that are hard-wired with the device they are harbored in (e.g., a vehicle or a digital camera).

Specific computers that are used for running-nodes in communication networks (such as routers and switches) will be described in section 2.4.1.

Due to ongoing developments in the hardware sector, the following categories of computer classes are not enumerative and classify only the most typical features of each category.

## 2.3.1   Mainframe Computers

*Mainframe computers* offer an extremely high processing speed within the multi-using mode. Usually, they are located in an air conditioned and otherwise protected computing center with various safeguards (e.g., in case of fire) and are managed and serviced by specially trained operators.

Mainframe computers regularly have extensive storage capacity. Making sure that the extensive administrative tasks (e.g., the controlling of the usage of the central processing unit) does not become a burden of the host (as otherwise considerable and valuable computing power would be diverted), one tends to work with powerful control units or front-end processors (e.g., IBM 3745). Moreover, in larger enterprises people often connect several hosts within a network in order to, e.g., meet high performance demands of users or to maintain a certain fail-soft service in case of system outages (see section 2.4.2).

New installations of mainframe computers can be regarded as questionable, when, e.g., PC networks may offer almost similar performance. Mainframes are usually not very user-friendly in comparison to PCs. Therefore, the host as an immediate partner in a man-machine dialog has an inferior position. Enterprises strive to take suitable applications away from hosts and tend to implement these on smaller systems (downsizing) in order to gain cost advantages, e.g., when utilizing workstations. An example is the displacement of the SAP R/2 system (designed for the mainframe) with SAP R/3 which was designed for Unix, as well as Windows NT. Nevertheless the host in business information systems is still being used as a "coordination authority" in networks and as a place in which central data are being stored. It is often the center of a star network (see section 2.4.2) connecting many PCs.

## 2.3.2   Workstations

In principle *workstations* are conceived as stand-alone computers in a workplace whose performance capacity is below that of mainframes. In particular they are useful in the support of calculation-intensive tasks in technical and scientific applications such as with computer-supported design (see section 5.1.1.1). These applications are less suited for mainframe computers as they demand a constant and high computing performance that is not permanently available if many users access a host in a parallel fashion. Additional performance may be achieved when workstations are equipped with several processors. Moreover, *workstation farms* may facilitate a demand distribution by utilizing momentarily unused capacity (see section 2.4 and 2.5.5).

A workstation is in its external dimensions barely larger than PC housings and may be stored in places that save work space (e.g., on or under the desktop, in a computer cabinet). Workstations are typically networked and equipped with a UNIX or Windows NT operating system, but also Linux is increasingly used (see section 2.4.2). Alternatively, high-performance workstations also are applied as departmental computers (i.e. as "small" central computing devices at the departmental level that are utilized as the center of a star network at the departmental level or for mid-sized enterprises). In large companies these units are often connected to the enterprise-wide central computers. Workstations may be operated in "normal" offices and can serve from about 20 to 200 terminals or PCs. Often numerous workstations are installed in special rooms equipped with security and air conditioning safe-guards.

### 2.3.3    Network Computers and Thin-Clients

As an additional class of computers one often discusses network computers (NC) and thin-clients. These are low-priced computers with low performance especially conceived for client operation in networks (see section 2.4.4). The fundamental idea is that application systems run on a distant server and the NC or thin-client receives data via the network and displays it. Conversely, the NC or thin-client sends input data to the server. In an ideal case such a system works without having a hard disk. Through central administration (e.g., in a computer center) the costs for the maintenance of such a system may be reduced.

## 2.4    Computer Networks and Network Architectures

So far the single computer was given focal attention in our discussion. Following we will address the integration of computers in closed networks, i.e. those whose components are under the control of an operator. Networks are the fundamental requisite of decentralized application concepts. With the deployment of computer networks the following *goals* are being achieved:

- Load balancing/performance pooling
- Data integration
- Program integration
- Communication integration
- Device and security integration

*Load balancing* leads to a better utilization of the capacities of computers on the network as the least utilized computer is ready to perform a new task. Beyond that, *performance balancing* aims to let an extensive task, which cannot be worked on efficiently by an individual computer any longer, to be processed in a parallel fashion in a network of several machines at the same time.

*Data integration* permits parallel access to data available in the network by several available computers or respective users.

*Program integration* allows the common use of a program by all computers connected within in the network. Program integration helps to avoid multiple procurement and parallel maintenance efforts. The program must, however, apart from exceptions, be network-ready. In addition, network licenses are needed that commonly have a limit on the number of concurrently signed on users.

*Communication integration* provides the basis that, for example, users of different computers may message each other. The messages themselves may be stored within an electronic mailbox.

Through *device integration* it is possible for all computers on the network to get access to all resources. From a cost perspective the joint usage of expensive peripheral devices such as high-quality laser printers is efficient.

A *security integration* is designed to guarantee the access to critical data via multiple paths (e.g., the parallel storage of the same data on two different computers). In the case of technical problems one may be routed to the alternative access path.

Computers are networked to integrate decision makers (people or machines) into joint, but distributed disposition or planning processes. Examples are varying forms of inter-organizational integration (e.g., electronic data interchange (EDI) in the area of *supply chain management* (SCM) in section 5.4) or the usage of external data bases (e.g., during patent enquiries, see section 3.2.4).

In the next section we will first address network components, followed by explanations about local area networks and wide area networks that have been constructed using these components. Finally we regard the contemporarily central cooperation model among computers, the so-called client-server concept.

## 2.4.1   Components of Computer Networks

When independent computers are connected via communication paths to exchange information, we may denote that a *computer network*. Networks can be connected to other networks and be comprised of several *sub-networks*. The most important components of a computer network are:

- The computers themselves, including their physical network connection (network card or modem), as well as the respective operating, network and application software

- Connection and communication computers in and between networks (hubs, routers, switches, bridges)

- The data transmission paths

- Protocols (see section 2.2.1.4)

*Computers* must possess either a network-capable operating system (e.g., UNIX) or specific *network software* when they shall run in networks. Well known products in the area of PC networks are, e.g., NetWare by Novell™, as well as MS Windows™ NT.

Connection and communication computers are specific devices whose task is the integration of computers in networks, the connection of networks, as well as the intelligent transmission of data packets. Often they are referred to as switch stations or as transmission hubs.

*Hubs* are central points within a local area network (see section 2.4.2) that connect computers. The connection occurs via a series of connectors that are referred to as ports. Data packets arriving in ports are being copied within the hub and are, subsequently, being transmitted to all other ports and thus computers as well.

Two local networks of the same type may be interconnected via a *bridge*. Bridges transmit data packets usually without any analysis of content and limitations to a recipient network. Aside from the transport of packets, so-called *switches* undertake also the function of data filtering. *Routers* connect differing LAN types. They are capable to react flexibly to errors in a complex connection situation and certain load conditions, and, to the degree to which suitable resources are available, they may identify alternative routing paths utilizing other routers. This is partially made possible through the application of optimization algorithms for the various data packets to determine the momentarily best possible path.

Through selective combinations of these concepts one attempts to combine varying advantages of the individual devices. Accordingly, one utilizes *routing switches, switching hubs* or *bridge routers (brouters)*.

The data transmission between connected computers takes place via *data transmission paths* (lines or radio links). The most common cable types are:

- Twisted pair copper wire

- Coaxial cable

- Fiber optic cable

*Twisted pair copper wires* are a very wide-spread transmission medium. The advantage of an easy installation of copper wires is accompanied by the disadvantages of electrical interference sensitivity and wire-tapping insecurity. Such copper wires are the cheapest transmission medium.

*Coaxial cable* is being used as well for television cable or antenna connections. It is shielded, enjoys improved performance characteristics and is less interference-prone than twisted pair copper wire. In comparison, the material is more expensive and the installation is more difficult.

*Fiber optic cable* permits very high transmission speeds. Moreover, it is not very sensitive to interference, is wire-tapping safe and, judged by its performance characteristics, relatively affordable.

Increasingly, data are also being transmitted via *mobile or optical point-to-point radio systems*. To bridge large distances, also when using mobile applications, one typically chooses satellite-based transmission. With optical point-to-point radio systems infra-red light or laser light transmissions are used. They are implemented to bridge distances up to 5 kilometers.

Based on the above described transmission layers, the most commonly used protocols (see section 2.2.1.4) are:

■ TCP/IP (Transmission Protocol/Internet Protocol), used for Internet transmissions (see section 2.5.1)

■ NetBEUI (NetBios Enhanced User Interface), developed for Windows-based networks

## 2.4.2 Local Area Networks

Locally connected computers in the same office, house or same office complex are referred to as *Local Area Networks (LAN)*. These are often operated by local network departments within companies. Technically the maximal distance between two computers may not be more than a few hundred yards.

In local area networks that are not cable-bound, i.e. *Wireless Local Area Networks (WLAN)*, mobile devices such as laptops may communicate among each other via infrared or radio technology using wall or ceiling-mounted access points within that network.

Based on the nature of the connection, Local Area Networks may be differentiated into four topologies, (Figure 2.4.2/1):

■ The ring network, especially IBM's token-ring network

■ The bus network, especially the Ethernet concept (Xerox™)

■ The star network

■ The intermeshed network

In a *ring network* computers are connected logically in a circle. All machines participating in the network have equal priorities and access rights. Data are being transmitted in one direction only. Such a configuration has the advantage that data routine is simple. On the other hand, the breakdown of one computer results in the breakdown of the entire network. In an effort to avoid this danger the applied physical typology often is a star network. Then, a failure of a peripheral computer effects only a limited performance restriction for the network.

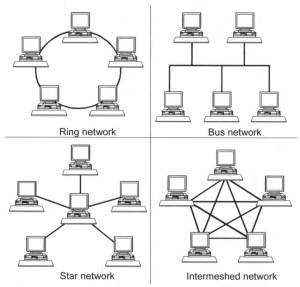

*Fig. 2.4.2/1      Important Typologies of Local Area Networks*

For the control of transmission and reception operations in the logical ring one utilizes the *token procedure*. Here, a token—to be envisioned analogous to a baton during a track relay race—circulates in the network automatically. Technically we may view a token as a defined bit sequence. A station that would like to transmit data to another station waits until the "empty" token has arrived and it may transmit data only at that moment in time, i.e. by occupying the empty token.

In a *bus network* all computers are connected to a common bus cable. Before transmitting each computer checks if the bus is free and then transmits into the bus. Collisions may occur anyway, i.e. in the case that two or more machines transmit simultaneously. Since each station is capable of listening to the medium while sending its own transmission, such conflicts get recognized, a so-called "jamming" signal is being transmitted and the transmission is retracted. Not until after a certain delay (for each station this delay may be different) a renewed transmission is being attempted. This method is referred to as the *Carrier Sense Multiple Access with Collision Detection* (CSMA/CD).

In contrast to these PC-LAN applications, designed on the basis of autonomous machines, mainframe-oriented LANs (see section 2.3.1) deploy typically a *star configuration* with the host in the center. The star typology has the disadvantage that the entire network collapses if the host fails.

When additional connections are made to a star configuration between peripheral computers, one refers to a partially *intermeshed network*. If each computer is directly connected with each other computer, we refer to a fully intermeshed network. Such a network structure has the advantage of being very effective in terms of performance. On the other hand, installing cable is time-consuming and thus expensive.

## 2.4.3 Wide Area Networks

Local computers or networks that are geographically far apart may be connected via *Wide Area Networks* (WANs). We may distinguish between closed WANs, e.g., via secure procedures (see section 2.5.4) intended for defined user groups and the public WAN—the Internet (see section 2.5). As a technical infrastructure one utilizes cable and wireless transmissions within various *network services*.

- Telephone network

- Direct dial network

- Integrated Services Digital Network (ISDN)

- Asymmetric Digital Subscriber Line (ADSL)

- Wireless networks

The *telephone network* is a network with dial-up capability and is widespread through the world. A disadvantage is the relatively high error rate throughout noise, as well as the relatively slow transmission capability. A microcomputer is connected to the network using a *modem*. The modem's task is to convert the data to be transmitted into analogue signals such that they become transportable via the telephone network. Costs for the use of the telephone network vary (aside from fixed basic fees), since these are calculated depending on duration and time of day for data transmissions or the distance to be covered, respectively.

A *direct dial network* is comprised of a dedicated line between two or more computer nodes. In contrast to the telephone network situation, here only fixed costs arise in reference to a contractual time frame, e.g., three months or a year.

The *Integrated Services Digital Network* (ISDN) is based on a digital network enabling sender and receiver to transmit information in varying forms (text, data, language, graphics). The network moreover enables the integration of the offered communication services, as phone conversations, fax and data transmission are carried out using a uniform number. A requirement for the implementation of ISDN is that the analogue telephone network is digitalized. ISDN offers aside from flow rate advantages a secure transmission of information over long distances. Moreover, it is possible to hold a separate telephone conversation while simultaneously being connected via other channels or to send or receive a fax.

*Asynchronous Transfer Mode* (ATM) is the broadband data transfer technology for ISDN. When using ATM, data streams are being dissected into packets of a constant size. Their transmission occurs while a connection-specific, logical channel is assigned for each partial distance between two connection points. For each transmission segment a fixed physical route is reserved. In the future billing for transmission services is to be simplified through this application principle.

*Asymmetric Digital Subscriber Line* (ADSL) is a high capacity data transmission that when compared to ISDN permits maximum transmission speeds 64 times faster between the switching exchange and the user installation (PC, workstation) without having to make any changes in the cable or wiring infrastructure. ADSL offers a high-speed data transmission rate to the user, yet a lower transmission speed in the opposite direction. The data transmission occurs in a frequency band above the voice channel such that the telephone/ISDN operation is not curtailed.

Aside from the above described *terrestrial networks* we need to consider *mobile phone networks*. Various service providers offer state-wide or regional coverage, although in the United States and Canada frequently less populated, and especially remote areas are not covered. Cooperation among competing service providers enable the user's access to other networks. In the European Union efforts are afoot to introduce the future mobile phone standard UMTS (Universal Mobile Telecommunication System) to replace step-by-step GSM (Global System for Mobile Communication) technology.

Of particular importance are wide area networks in conjunction with *mobile computing*. The capture of data occurs closer to the place of origin such as in the situation when a field service employee of an insurance company inputs data at the customer's location (see section 5.1.2.1). Such data may then be transmitted simultaneously or relayed over WANs to a central host.

A *backbone* is a central transmission route within a network that bundles data from various connections and sub-networks before transmitting them via high performance devices. At the local level a backbone connects nodes that bring together geographically distant LANs. Within the Internet, backbones are those sub-networks or connections that bring together local or regional networks over large distances.

## 2.4.4    Client-Server Concept as a Model for Cooperation

Communication among computers presupposes the existence of a suitable model of cooperation. This, in turn, determines a definite role distribution and specifies common protocols. In the *client-server concept* on the user-side so-called clients utilize certain services (e.g., data and transactions of an application system) provided by a particular computer in the network (server). The task of the client is the presentation of corresponding data and the interaction with the user. The server, e.g., a data base server, waits passively until a query is received from the client. Within the network one or more computers can be designated as servers (cf., figure 2.4.4/1).

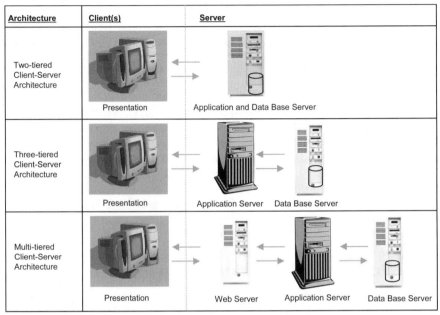

*Fig. 2.4.4/1     Client-Server Architectures*

*Integrated* standard software solutions (see section 2.2.2.1.3), e.g., SAP R/3 or Baan™ IV, may be subdivided into three different functional tiers or layers:

- Presentation (preparation of the graphical platform)
- Application (provision of the application logic)
- Data management (administration of data)

SAP R/3 offers based on the client-server concept, e.g., the possibility to distribute the various layers to different computers within a network, decoupling the application logic from presentation and data management.

With two-tiered configurations application and data base services are being implemented on one server. In the three-tiered case a separate computer is available for application and database services. In order to achieve load balancing on the applications level, different application servers may be operated simultaneously, e.g., for sales and financial accounting.

In order to enable web-based access for applications a web server may be implemented that functions as the intermediary between presentation and application level. The resulting architecture is labeled as four-tiered or multitiered. Web servers are among the most important connecting links in the

Internet. Based on the query by the client (in this case a web browser) they transmit the requested HTML pages and enable the generation of dynamic documents over interfaces, These documents are send back to the user afterwards as a result of the execution of programs.

An important criterion for the decision for a specific (e.g., two or three-tiered) architecture is the number of users who work with the application in a parallel fashion. A distribution to different systems offers a better performance that may be raised through scaling (e.g., the installation of additional web and application servers).

The structure of this model of cooperation presupposes an operating system on the server side that permits multi-tasking (e.g., UNIX or Windows NT). Otherwise, the entire computing performance of the machine would be blocked through the permanently running server process. In large networks different computers serve as both clients and servers. This then is referred to as *peer-to-peer communication* (communication among equals). For the time being, all important services in the Internet (see section 2.5.2) are based on the client-server concept.

## 2.5 Worldwide Networking: The Internet

The term *Internet* refers to the combination of thousands of local area networks consisting of millions of computers (cf., fig. 2.5/1) that exchange information via the protocol family TCP/IP (see section 2.5.1). Moreover, it offers a set of services and techniques that are not only important for its own functioning, but also offer a multitude of impulses for network users.

The development of the Internet is characterized by the attempt to provide access to information resources at any time at an affordable price in order to establish cooperative advantages by the worldwide connecting of networks. The idealized goal is that all LAN and WAN operators make available parts of their closed-network infrastructure within the Internet. In doing so, each of them partly runs the meshed, decentralized and thus largely failure-safe public network.

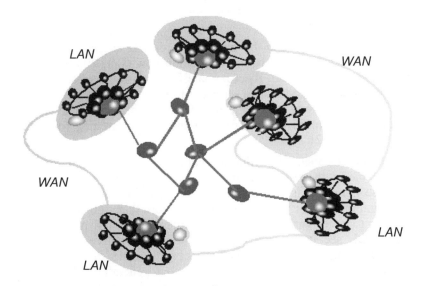

*Fig.2.5/1      Architecture of the Internet*

The fundamental motivation of local area network *providers* is based on the expectation that one's own "customers" find additional value in the worldwide offering of information to cover the costs of the appropriation. The calculation of wide area network providers is based on two arguments: On the one hand Internet services contribute to a leveling of the network load so that the connection infrastructure is better utilized. On the other hand, providers hope for the fast expansion of their business together with an equally rapid diffusion of sales of various value-added services.

Information and communication services connected to the Internet are generally perceived as promising growth areas for the future. Offerings range from simple data transmissions (e.g., digital pictures and music data bases) to online shopping on electronic marketplaces.

Based on figures by the EITO Observatory (2002) and Koenig, Wigand and Beck (2002) there are 169.5 million Internet users in the United States. Few technologies have spread as quickly, or become so widely used, as computers and the Internet. These information technologies are rapidly becoming common fixtures of modern social and economic life, opening opportunities and new avenues for many businesses and consumers alike. According to a National Telecommunication & Information Administration report [NTIA 02], the rate of growth of Internet use in the United States is currently two million new Internet users per month. In September 2001, 174 million people (or 66 percent of the population) in the United States used computers. Children and teenagers use computers and the Internet more than any other age group. Internet use is increasing for people regardless of income, education, age, race, ethnicity, or gender.

Americans are going onto the Internet to conduct an expanding range of activities: Forty-five percent of the population now (Oct, 2002) uses e-mail, up from 35 percent in 2000. Approximately one-third of Americans use the Internet to search for product and service information (36 percent, up from 26 percent in 2000). Among Internet users, 39 percent of individuals are making online purchases and 35 percent of individuals are searching for health information [NTIA, 02].

For the long distance traffic the Internet relays networks that are typically made available by the telecommunication service providers. Thus the capacity of those networks is of considerable importance. With *Internet 2* (see http://www.internet2.edu), an undertaking of more than 160 North American universities together with the U.S. Government-sponsored *Next Generation Internet* (NGI, see http://www.ngi.gov), various initiatives are being discussed addressing the further development with regard to capacity bottlenecks and improvements of services. The next generation of an improved Internet infrastructure will enable the widespread usage of data extensive multimedia applications such as telemedicine, digital libraries or virtual laboratories. Moreover, the coalescence of television and Internet infrastructure will be observable in the future.

In the following paragraphs we first describe the protocol family TCP/IP, followed by the Internet services based on these protocols. A discussion of the configuration of private networks by utilizing Internet technologies in intranets and extranets follows. Finally, we will explain security techniques for public and private networks.

## 2.5.1    The Protocol Family TCP/IP

The protocol family *TCP/IP* is comprised of two parts, the *Transmission Control Protocol* (TCP) and the *Internet Protocol* (IP). TCP dissects information, e.g., electronic mail, into different data packets and assigns each data packet with an IP address of the sender and the receiver.

*IP addresses* are cipher codes to identify information locations. They have a length of 32 bit (= 4 byte) and are represented by four decimal numbers separated by periods or dots. Each one may take on the whole-number value of the interval between 0 and 255. For people it is usually easier to deal with names than number columns. For this reason a logical name structure exists that assigns each computer a hierarchically structured designation. The so-called domain name service (DNS) translates this name (e.g., for the University of Arkansas at Little Rock it is: www.ualr.edu) into the corresponding IP address (144.167.10.128).

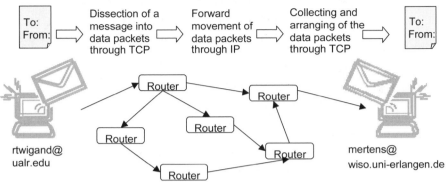

*Fig. 2.5.1/1    Data Transmission over the Internet*

Figure 2.5.1/1 exhibits an exemplary data transmission over the Internet. The packets are sent to a *router* (e.g., to the router of the Internet provider) whose task is to forward the information within the IP-controlled transmission. Within the network of routers, telephone companies attempt to fill out momentary low load demands of parts of the infrastructure. This is achieved by sending the packet via a path that is least utilized on the way to the destination. This also explains the reason for the low transmission costs over the Internet. Aside from this advantage, the Internet does not prescribe any fixed routing paths. Each data packet of a message may be routed another way in the Internet, i.e. within the packet switching network. At the destination their respective rearrangement into its original form is made possible through TCP.

The continuous expansion and installation of new transmission routes and the increased meshing of the infrastructural capacity prevent bottlenecks. A problem exists with regard to the addressing, even though theoretically about 4 billion systems may be addressed. Closed networks have at their disposal a more or less large address space within which several subsystems may be operated, while often large portions of the address space remain unused. Thus Internet addresses become scarce.

A new protocol conversion, i.e. *IPv6*, is supposed to replace the 32-bit version of IP in a stepwise fashion. With version 6 128 bits are used for the addressing which corresponds to a quantity of $3.4 * 10^{38}$ of addresses. While the previous address numbers were represented in decimal numbers separated by periods, addresses using the IPv6 format are represented by eight hexadecimal numbers (e.g., 2BA:0:66:899:0:0:459:AC39) and separations occur using colons [Hagen 02]. Aside from the enlargement of space for addresses IPv6 is to simplify routing and should enable greater data security (see chapter 2.5.4).

## 2.5.2    Services and Technologies of Networking

The Internet offers today a multitude of services that make it possible for the user to receive or to send information. Among the most popular services is the *World Wide Web* (WWW).

A central building block of web applications are documents based on the Hypertext Markup Language (HTML) (see section 2.2.2.1.1). Since older versions of HTML offer only simple functionalities for the designing of web front-ends, control features such as pull-down menus (well known from graphical user interfaces (GUI) programming) with embedded scripts in languages such as JavaScript need to be programmed subsequently. However, based on XML the current xHTML version also offers interactive features.

Due to the far and wide success of the WWW the limitations of the HTML concept have become visible, as, e.g., the object structure of the output data is not explicit and their further processing is made difficult. A solution to this problem is offered by the Extensible Markup Language (XML) [Hasselbring 00]. This meta-language opens the opportunity to describe data in the network such that the underlying data structure can be transmitted to distant users and their information processing systems. It is therefore, e.g., possible to exchange varying standardized commercial documents among different systems as a new form of Internet-based Electronic Data Interchange (EDI).

A substantial expansion of web applications has been triggered by the Java programming language and Java technologies (e.g., JavaBeans or Enterprise JavaBeans). Servlets, e.g., are special server programs written in Java that are executed on a web server without any classes for the depiction on the web front-end. These applications may be initiated by a client through a special query to the web server.

HTTP (*Hypertext Transfer Protocol*) describes the standard protocol of the WWW. The hypertext connection enables user-friendly access to distributed information on the Internet. This is made possible by clicking a specially marked link that, in turn, evokes the display of another document. Each web management system contains in addition to the provided documents a so-called HTTP demon that is waiting for HTTP queries and that will answer these. The web browser is an HTTP client that generates queries and sends these to the web management system, which processes these and returns the desired document. Since HTTP was conceived originally only for the free exchange of scientific publications, such connections may be tapped with little effort. In the commercial evolution of the Internet HTTP security features were added later on [Totty 02].

CGI (*Common Gateway Interface*) is a component of the HTTP specification and represents a standardized interface in order to address external programs from a web management system. When the user completes an HTML form the web server initiates an external program that processes the data and generates the answer. Users, e.g., are offered the possibility to access data (see chapter 3) bases via the web. In order to enable queries from a browser to a server and to provide an answer to the query, a web server is activated that transmits the received parameters from the client via the CGI to the server. The server, in turn, directs the answer back to the client using the very same path. CGI-based architectures established themselves in the meantime as a quasi standard for Internet applications, even though their range of application due to the limited scalability is limited. Another web server-based script becoming increasingly important is PHP (Hypertext Preprocessor) [Williams 02].

Aside from the WWW other application-related services exist that are based on TCP/IP. Such major systems are:

- Electronic mail (see section 2.2.2.1.1)
- FTP (File Transfer Protocol) for data transmissions
- USENET to participate in theme-specific discussion
- IRC (Internet Relay Chat) for the conduct of online conferences
- Telnet (remote access for virtual terminal emulation)

  The Telnet service allows users—if authorized—access from their computer via the network to services of another computer seeing on their screen the services of the other computer.

- Voice over IP

  IP-based telephone services referred to as *Voice over IP* are of increasing importance. Voice information is gathered via a microphone and digitalized through the use of a soundcard. It is then dissected into IP-packets and transmitted via the network like common data. When the packets arrive at the destination (e.g., a telephone extension) they are reassembled into their original sequence.

  Through packet switching within the Internet the individual parts of a message may travel varying paths, such that through slow delivery of a data packet delays are possible when the voice is being put out again. Support for such shortcomings is being promised by IPv6 (see section 2.5.1). This protocol version offers the possibility, just like in the switched network, to establish a predefined, fixed connection.

## 2.5.3    Intranets and Extranets

The described Internet technologies, as well as the often freely available software become increasingly interesting for the broad-based deployment within an enterprise's non-public networks such as Intranets or Extranets.

*Intranets* are closed networks based on TCP/IP and corresponding protocols and services. The design of Intranets is especially attractive for integration purposes with public Internet services to utilize both networks with the same platform. Often internal handbooks, circulars, address directories, organizational directives and non-public parts catalogues are available in Intranets. Firewalls are often implemented as interfaces between the closed network and the Internet. These partition the internal security zone from the public network, by controlling all incoming and outgoing data packets and refuse passage to unauthorized packets.

An *Extranet* on the other hand is a closed network available to firms connected via the Internet that have specific access rights (e.g., suppliers of an automobile manufacturer). Just like Intranets, Extranets are based on the usage of Internet protocols. A widely used technology are so-called *Virtual Private Networks* (VPN) in which the *tunneling protocol* encodes information in the transition from a private LAN into the public network (more correctly: within the firewall). The encrypted data have to be decoded at the destination point accordingly. This technology may also be utilized within Intranets.

A large VPN is ANX™ (Automotive Network Exchange, http://www.anx.com) of the United States automobile industry. Aside from three large manufacturers (DaimlerChrysler™, General Motors™, Ford™) it connects more than 10,000 suppliers and 40,000 dealers.

## 2.5.4    Security in Information and Communication Networks

Just as in normal life absolute *security* is equally impossible to achieve within information and communication networks.

Security problems within a *closed network* occur, e.g., when the log-in of unauthorized individuals cannot be prevented or when users behave improperly. Three types of access assurances are being deployed in order to avoid this [Buxmann 00]:

- Verification of person-specific criteria, e.g., by the use of finger prints or facial recognition systems

- Verification of bearer-specific criteria based on hardware, e.g., by the use of a smartcard

- Verification of bearer-specific criteria based on software, e.g., by the use of a password or a PIN (Personal Identification Number)

Hardware-based access protection that captures unalterable personal criteria offers the highest degree of security, but is also the most expensive solution. Software-based solutions are affordable, but of limited reliability. In practice, mechanisms are often utilized that are combinations of bearer-specific criteria examined through hardware and software (e.g., the verification by card and PIN).

In order to avoid errors for authorized users it is possible to specify access paths via the operating system considering programs as well as data. In that way a usage guide is established that follows specific rules.

Security problems in the *public network* result when many users without initial screening and verification utilize the largely unregulated medium Internet. From the perspective of the user, an increased effort to assure security stays in contrast to the relatively low costs for connection and data transmission. Within the Internet, as well as within Intranets and Extranets, fundamentally three types of attacks are possible:

- Spy attacks or data sniffing by unauthorized users to intercept or to change a communication between actors

- Denial of services attacks to paralyze servers or attempts to change their contents

- Virus attacks as non-directed attempts by users in the network to prevent communication or to delete files

In an effort to assure secrecy and authentication cryptographic procedures may be used. One may differentiate between *symmetrical* and *asymmetrical encryption procedures*.

In symmetrical encryption procedures a message is encoded by the sender using a key and it is being decoded by the receiver on the basis of the same key in inverted direction. Keys are mathematical algorithms coding, e.g., text into cipher sequences. One problem, however, is that first both sender and receiver must exchange the key via a secure channel.

The basis of asymmetrical deciphering (avoiding the aforementioned problem) is the creation of a so-called *public key* that is published, e.g., on a home page and a closely related *private key* that, e.g., may be stored on a smartcard. The generation of the keys may be initiated by one of the participating actors (sender or receiver) or a so-called *Trusted Third Party*, a neutral third person or institution (cf., fig. 2.5.4/1). The sender utilizes for the encryption of a message the public key of the receiver. The thus encrypted message can only be decoded by the receiver with his/her (only known to him/her) private key.

Symmetric and asymmetric procedures are often combined. An example is the procedure underlying the *Secure Socket Layer* (SSL) that is applied when using, e.g., a web browser. At the beginning of a session one first generates symmetric keys that are being exchanged among the participants using asymmetric encryption. Following a transmission of the message occurs with symmetric encryption.

Beyond this, there is a need to prove that the transmitted message arrived safely and "untouched". This occurs with so-called hash functions, i.e. by adding error-checking-numbers created by bit operations to the document which serves as an integrity certificate. The recipient of the message recreates the error-checking number using the same procedure and compares both.

Asymmetric encryption and hash functions are combined in the creation of a *digital signature*. Figure 2.5.4/1 illustrates the cooperation of both procedures.

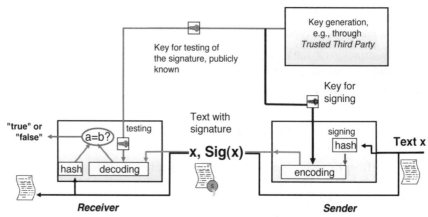

*Fig. 2.5.4/1     Creation and Verification of a Digital Signature*

The sender generates the digital signature by creating the hash value of the text $x$ which is encoded with his/her private key (*Sig (x)*) and sends it with the original text ($x$) to a receiver. The receiver who knows the public key of the sender decodes the hash value ($b$) and generates via the original text the object of comparison ($a$).

During attacks on the operational readiness of servers, e.g., a web server (see section 2.4.4) may be confronted with a very large number of artificially created queries that contribute to the crash of the computer or the overloading of lines (denial of service attack). Since one cannot easily distinguish between valid and useless queries, security or protective programs offer little help. Attacks whose goal it is to alter the content on a server may be countered by turning off unnecessary services and the installation of appropriate security software.

Regarding untargeted attempts to disrupt communication, the generation and diffusion of viruses are of central concern. For protection one deploys virus detection programs.

## 2.5.5 Computer and Network Infrastructures

Enterprises and other organizations offering network services utilize the here described hardware and software building blocks to configure their computers and network infrastructure and connect these with the Internet. In large firms this infrastructure development occurs during the course of various acquisition and expansion decisions, as well as in conjunction with the increasing integration of business management and technology. Often this development starts with centralized mainframe computers while a decentralized client/server environment is evolving. Unplanned short-term effects may lead to a situation in which such system structures grow in an uncoordinated fashion. A contribution to a better targeted development may be the application of *information systems architecture models* (see section 7.1.2). An additional development trend is the increasing integration into the Internet and the adoption of seasoned Internet technologies and products into one's own application on the basis of PC and workstation-based systems.

*A PRACTICAL EXAMPLE*

*Comdirect™ bank AG, a subsidiary of Commerzbank™ AG, is with 650,000 customers one of the largest direct-brokers in Germany. Of central concern is the quick, reliable and economical order execution for the buying and selling of stocks. 90% of all orders are executed via the Internet. For this purpose comdirect bank runs the online transaction system Comhome through which customers may place their orders. This system is connected via dedicated lines with the central computing system of Commerzbank AG, e.g., to consolidate accounts (cf. fig. 2.5.5/1).*

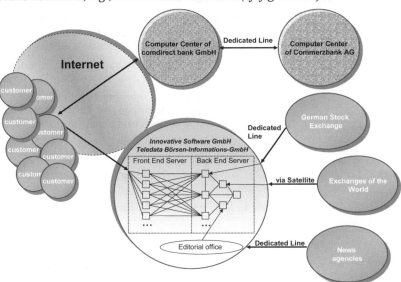

Fig. 2.5.5/1     *Excerpts from the Computer and Network Architecture for Comdirect bank AG*

*In addition to the Comhome system comdirect provides its clients access to the largest European stock exchange information service on the Internet. This service is called Informer and was developed by IS Innovative Software™ AG in Frankfurt, Germany (see* http://www.comdirect.de*). This system offers, e.g., free live information about worldwide stock exchange business and different platforms for beginners and professionals. During the third quarter 2000 Informer served about 250 million page impressions and about 55 million visits per month via the Internet. During peak demand times (shortly after the Frankfurt stock exchange opens and right before it closes) the system answers about 100,000 data queries per minute.*

*To meet such demands IS Innovative Systems AG is connected to the Internet via quadruplicate constructed 155 Mbit ATM lines. At peak time, Comdirect bank's capacity contingent is utilized at about 50%. Transaction performance is provided by a demand-scalable set of 165 Intel-based high-performance back-end servers that also communicate via broadband connections. Users utilize front-end servers that maintain the application logic and generate web pages (e.g., standardized stock price queries or individual market overviews). Back-end servers, providing customers indirectly (via the front-end servers) with additional information, hold historic stock price data bases and news archives. In addition they support intraday trade. The application software runs under Linux and is fully developed by IS Innovative Software AG in order to be able to handle the high volume of flow rate demands. Since the services Comhome and Informer are integrated in the web browser the user is unable to recognize the separation between the two systems.*

## 2.6   Literature for Chapter 2

| | |
|---|---|
| Bremer 96 | Bremer, M., Modales natürliches Schließen: modale Prädikatenlogik mit Existenz und Kennzeichnungen, aber ohne Possibilia, Darmstadt, Germany, 1996. |
| Buchmann 99 | Buchmann, J., Introduction to Cryptography, Springer Verlag, New York, 2001. |
| Buxmann 00 | Buxmann, P., König, W., Inter-organizational Cooperation with SAP Systems. Perspectives on Logistics and Service Management (SAP Excellence), Berlin, Germany, 2000. |
| EITO 02 | EITO Observatory (2002): Annual Report 2002. Frankfurt, Germany: European Information Technology Observatory. |
| Giloi 93 | Giloi, W.K., Rechnerarchitektur, 2nd edition, Berlin, Germany, 1993. |
| Hagen 02 | Hagen, S., IPv6 Essentials, O'Reilly UK, 2002. |
| Hasselbring 00 | Hasselbring, W., Information System Integration. In: Communications of the ACM 43 (2000) 6, pp. 32-38. |
| Koenig 02 | Koenig, W., Wigand, R. T. and Beck, R. Globalization and Electronic Commerce, in: Communications of the Association for Information Systems , Volume 9, Atlanta, October, 2002. |

| | |
|---|---|
| Messerschmitt 99 | Messerschmitt, D.G., Networked Applications, A Guide to the New Computing Infrastructure, San Francisco 1999. |
| NTIA 02 | National Telecommunications & Information Administration (NTIA) and the Economics and Statistics Administration: A Nation Online: How Americans Are Expanding Their Use Of The Internet. Washington, DC: U.S. Department of Commerce, 2002. |
| Scheer 94 | Scheer, A.-W., CIM – Computer Integrated Manufacturing. Towards the Factory of the Future, 4. Edition, Berlin, Germany, 1994. |
| Shapiro et al. 98 | Shapiro, C., Varian, H.R., Information Rules: A Strategic Guide to the Network Economy, Boston (Mass.) 1998. |
| Totty et al. 02 | Totty, B., Gourley, D., Sayer, M., Aggarwal, A., Reddy, S.: HTTP: the Definitive Guide, O'Reilly, UK, 2002. |
| Williams et al. 02 | Williams, H.E., Lane D., Web Database Applications with PHP and MySQL, O'Reilly UK, 2002. |

# 3  Data and their Integration

Decision making may be viewed as information processing. *Data* that are neutral with regard to purpose and intent (e.g., capacities, deadlines) and the derived purpose- and intent-directed *information* (e.g., a capacity bottle neck that may lead to target date delays during the processing of an order) serve as the basis for management *decisions* (e.g., the experience that specific data constellations trigger specific decisions). Data pertaining to enterprise-internal and external circumstances are thus the "raw material" for information and decision processes, as well as for the creation of knowledge.

Knowledge is always the result of processed information that is being interpreted within a context. Information is a signal derived from data, carries meaning and calls attention to something. Consequently, information leads to a change in existing knowledge. It follows then that data may be processed to become information that, in turn, may lead to knowledge.

Data may be encountered in all areas of an enterprise, as well as within its relevant environment and may relate, for example, to items and parts, orders, work places, consumption and payment processes, as well as characteristics and behavior of customers and suppliers. Such data need to be administered in accordance with the firm's strategy. This implies issues pertaining to data organization and these will be discussed in the first section on *data and databases*.

Enterprises usually are not limited to do their business in merely one location. It is no longer appropriate today to view a firm as a completed, integrated entity [Wigand et al. 97]. Modular organizations, cooperation models (such as, e.g., supply chain management, see section 5.4) and virtual organization structures pose new demands for data organization. Information is to be gathered today increasingly from data that are distributed worldwide. This theme is addressed in the second section *networked databases*.

The subject of this chapter concerns exclusively questions of data organization and data evaluation. We will not address data provision, a process preceding data organization and evaluation. Numerous examples in chapter 5 offer a multitude of insights regarding these topics. Hardware-related aspects of data capture, storage, output and transmission in computer centers have already been addressed in chapter 2.

# 3.1 Data and Databases

After a short description of the goals and the requirements for data integration the fundamentals of data organization (see section 3.1.2 to 3.1.3) are presented. A juxtaposition of data and database organization follows so as to view components of database systems in a constructive manner. Particular emphasis is given to relational and object-oriented database models (see section 3.1.8).

## 3.1.1 Goals and Requirements of Data Integration

Data in a firm are then integrated when technically similar data are organized only once. Each user or each application, respectively, relates to the same data basis. First it is assumed that this data organization occurs centrally. In section 3.2 this assumption is not applied.

The *goals* of operational data integration and expected *benefit increases* compared to isolated data handling are essentially:

- Improved information access for decision makers (for example, through the potential realization of data-intensive, organization- or application-spanning information systems)

- Streamlining of work processes (above all through the acceleration of information flow based on the removal of information flow restraints)

- Reduction of data redundancies (superfluous data repetition) in order to avoid inconsistencies (logical contradictions) and to lower memory costs (see section 3.1.4)

- Improvement of data integrity (correctness and completeness of data), e.g., through minimizing the pitfalls of manual data entry (see section 3.1.6)

- Reduction of the data capturing effort (e.g., based on the elimination of multiple data capturing of the same or similar data)

- Creation of the prerequisites for a functional and procedural integration of the data (see section 4.2).

To realize these goals—aside from adequate organizational conditions—*technical prerequisites* need most to be fulfilled (see section 2.4). It follows that data should be fully automated and captured early at the data source (e.g., orders while still at the customer's business, production data at the location of manufacture) and data flow should be possible without problems via computer networks that are suitable and compatible with one another.

## 3.1.2 Classification of Data

*Data* are here understood as machine-processed characters (fundamental elements of data representation) that describe objects and object relationships within the real world via their characteristics and thus represent these. For example, we may think of the data of the object 'item' (with the characteristics price, item description, etc.) and of the data about the relations of this object to other objects (e.g., customer, supplier).

Data may be classified, for example, according to the following criteria:

- *Type of character* or *data type*, respectively: Numerical (computable and processable numbers), alphabetical (letters of the alphabet) and alphanumerical data (digits, letters and special characters)

- *Form of appearance*: Linguistic (e.g., human spoken language, visual (e.g., graphics) and written data (e.g., text)

- *Formatting*: Formatted (e.g., format-bound tables) and unformatted data (e.g., form-free text)

- Position in *data processing*: Input and output data

- *Purpose in application (e.g.)*:
  - *Master data* that are rarely changed, e.g., personnel master data such as names and addresses
  - Inventory *transaction data*, e.g., warehouse in- and outflows
  - *Transfer data*, i.e. data produced by a program and then transferred to another program, such as data produced by a spreadsheet program that are then taken over by a presentation program
  - *Preliminary data* (open item), i.e. data that exist for a duration until a precisely definable event occurs, e.g., a payment

The annual balance sheet of a firm, e.g., contains numerical, written, formatted and output data relating to capital assets and floating capital, as well as equity capital and borrowed capital.

## 3.1.3 Basic Terminology and Data Organization

Important terms of data organization are (cf., fig. 3.1.3/1):

- *Data field* (data element): A data field or data element consists of one or more characters and is the smallest addressable, as well as interpretable data unit. A data field may be, e.g., an item number or an item name.

- *Data record*: Data fields that are related due to their content become (logical) data records. A simple data record for an item consists, e.g., of item number, item name, class of goods and item price.

■ *Data file:* All similar data records, i.e. all data records that are formally identical, are stored in a data file—in our example this is an item data file.

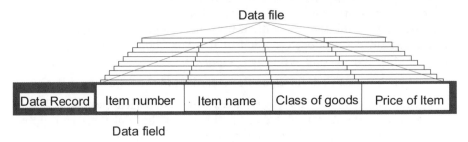

Data field

*Fig. 3.1.3/1     Hierarchy of Data Terms*

■ *Database*: A database is a data collection of data that logically belong together and that are stored in data files on a storage medium. For example, a simple database for cost accounting purposes may consist of the different cost elements, business cost centers and cost units.

■ *Database system*: A database system consists of a database and corresponding database software, the so-called database management system (see section 3.1.5).

■ *Distributed database system*: A database system whose database is not kept centrally in one location but in different locations and is connected via suitable networks is a distributed database system (see section 3.2.1).

■ *Data warehouse*: A data warehouse is a very complex database system whose data are fed by data from varying, internal and external databases and that has access to various tools for data preparation (see section 3.2.2).

■ *Internet*: The Internet represents with its innumerable, interconnected databases the largest existing data basis, which is at the same time not a database, as it does not utilize a database management system (see section 3.2.5).

In many databases data records have a *key*. A key consists of one or more data fields that identify explicitly a data record. Such identification must be possible independent of the data records momentarily present in the database. For example, if there is in a database at a particular point in time only one data record of the type EMPLOYEES whose data field SURNAME has the content "Meyer", then the data field SURNAME is not a key, as an additional employee with the surname "Meyer" may be added, i.e. hired, at any point in time.

Among all possible keys for a type of data record one key is marked, the so-called *primary key*. This key consists of the smallest possible number of data fields. An example of a primary key is the data field combination "Surname, First Name, Birth Date". If the "natural" primary key is comprised of a combination of key fields, then one often introduces for the purpose of easier identification of data records an additional data field as the so-called "artificial" primary key (e.g., "employee number" or "automobile identification number"). All keys that are not primary keys are labeled *secondary keys*.

Whether or not keys are assigned to data records depends essentially on the utilized database model. For example, keys are the lifeline for relational database models (see section 3.1.8.1), whereas with the object-oriented database model (see section 3.1.8.2) one may completely do without any sort of key.

## 3.1.4 Database Organization vs. Data File Organization

In the early days of data processing the development of AS was characterized by a *tight connection between program design and the physical data organization* in the storage media. In conventional programming efforts data are being made available on the data media in a program-related fashion. For each application, however, specific data files with the requisite data records, as well as specific functions are required. The definition of the needed data files occurs here in each respective application program. The data file design is adapted to the formulation of the task and has limited flexibility with regard to new applications, as available data files often need to follow a different sorting pattern or will have to be complemented with additional fields.

When managing data without databases this typically leads to a situation in which parts of the already existing data have to be created anew, thus it is possible that an uncontrolled *redundancy of data* comes into being. Data redundancy not only entails higher storage costs and increased expenditures with the documentation, but above all exacerbates the updating and securing of the data. In large AS we may encounter so many redundant data that it will be nearly impossible to keep all on an equally updated level. There will always be the danger of being confronted with inconsistent data.

In contrast to the data-oriented organization, the data of a database are of a general validity, i.e. they are independent of the individual programs by which they are used. This *independence* in the data organization places the essential demands on modern database systems. This is accomplished through the consequential separation into logical data structuring and physical data storage.

Figure 3.1.4/1 clarifies graphically the differences between data-oriented and database-oriented data organization [Ricardo 90, pp. 6 et sqq.]. In the first case programs 1 and 2 (and their users) have their own, physically present data files. Data file B is redundant. In the second case a database management system is available for use and the *logical data files* required in each case are available. Physically the data become redundancy-free and can be stored consistently in the database. The logical data files generally do not correspond directly to *physical data files*. They merely enable access to the data that the application program (and thus also the database user) needs.

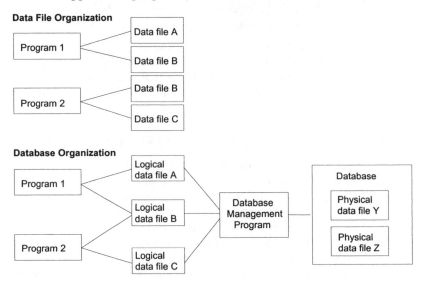

*Fig. 3.1.4/1     Data file and Database-oriented Data Organization*

In order to comprehend the data and database organization one has to understand data file operation and data access.

The following data file related operations (*data operations*) are possible:

- Searching from one or more data records for a specific search criterion (value of data fields)
- Changing data field values
- Entering new data records
- Erasing existing data records
- Sorting data records
- Copying of entire data files or parts thereof
- Partitioning of data files into several new data files
- Merging of several data files into a new data file

## 3.1.5  Components of a Database System

A *database system* (DBS) consists of a database (see section 3.1.3) and a corresponding database software, i.e. a *database management system* (DBMS).

The database management system takes on the administration of the database (processing of data operations) and provides at least the following features:

■ A data definition or description language (DDL = Data Definition/Description Language)

■ A data manipulation language (DML = Data Manipulation Language)

■ A data storage description language (DSDL = Data Storage Description Language)

The *DDL* is used in describing the logical data structures of a database. The *DML* enables database users and application programs to gain interactive access (e.g., changing, adding, deleting) to the database. The description of the physical data organization within a database system is accomplished by the *DSDL*. It makes possible the already described forms of the storage organization, but also comprises concepts going beyond this (above all the flexible linking of available data). These mentioned languages, however, are not always separated from each other in real DBS. Often a language comprises two or more of the components referred to above.

*Database query languages* (DQL = Data Query Language) simplify the direct communication between user and DBS and are not a compelling component of DBMS. These languages are mainly for the uncomplicated acquisition of information from large databases and require—in contrast to DML—no detailed system knowledge. In DBMS, however, the DML and the query language are often integrated into one unified concept. The de facto standard with relational DBS (see section 3.1.8.1) is at present the *Structured Query Language* (SQL, see section 3.1.9). Its object-oriented counterpart is the *Object Query Language* (OQL) for object-oriented databases.

## 3.1.6 Architecture of a Database System

When formulating data and data relationships one differentiates among three different levels of abstraction or perspectives: From an abstract or *conceptual* perspective, data and their linkages are formulated as situation- and thus also person- and context-independent as possible. From a second perspective, data may be organized in the way in which different users need them (*external view*). Finally data may be described regarding the structure of their physical storage (*internal view*). These three varying perspectives underlie the principal structure of DBS usually ascribed to the Three Level Architecture proposed by the ANSI/SPARC (American National Standards Institute/Standards Planning and Requirements Committee) (cf., fig. 3.1.6/1).

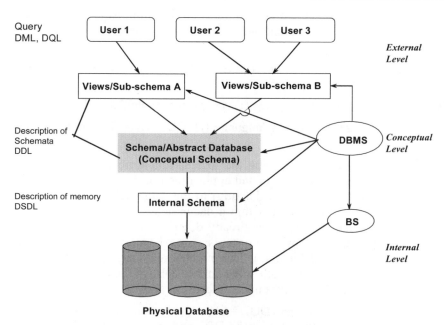

Fig. 3.1.6./1    *Three Level Architecture of Database Systems*

The *conceptual level* (also *conceptual schema*) comprises the complete logical description of all relevant objects and their relations. Conceptual models (e.g., the description of the data structures in materials management) are usually developed in cooperation with the departments within a firm. The DDL of a DBMS supports the transformation of the schema in the database description (especially the definition of data fields, field types, field lengths and the relations among data records).

The *external level*, represented through the sub-schemata, shows the description of data and their relations from the view of the individual user. The relation between the DBS and the users, as well as their application programs is created via the DML. The user's view of the utilized data records are called *sub-schema, external model* or *view* (e.g., a buyer is not permitted to see the monthly salaries of his colleagues; the human resources employee in charge of personnel salaries, on the other hand, requires these data). The problem- and user-specific views are derived from the conceptual models. Accordingly, the user view is an excerpt derived from the formal transformation of the conceptual model. Neither the schema nor the sub-schema specifies how data need to be stored physically. With a given logical data organization differing possibilities exist to organize the data.

In the *internal level* (also *internal schema*) the physical data organization is determined through the use of DSDL. The physical model provides a formal description how the data should be stored and how they may be accessed. This description is also referred to as the *internal model*.

The connections between the objects of the various levels are modelled by so-called *transformational rules*. These specify the way in which a specific type of a certain object of a model (consisting of one or more objects) may form a deeper, underlying model. The transformations between the individual levels are executed by the DBMS. It makes sure that requests for access are made by the terms of an external model. Moreover, the execution of the necessary operations on the physical level is enabled by the management system, and the desired data (in the form defined by the external model) will then be transmitted to the user.

Important features of DBS are:

- *Data independence:* Independence between schema and application programs or users (logical data independence), as well as between schema and physical data organization (physical data independence)

- *Planned and controlled data redundancy*: Limiting redundancy to a smallest possible or purposive degree, respectively

- *Assurance of data consistency*: Equal update level with multiple-stored data holdings

- *Database integrity*: Correctness and completeness of data. Avoidance of incorrect entries and impermissible operations (semantic integrity), as well as the avoidance of errors when multiple programs or users access the same data inventory (operational integrity)

- *Data security*: Ensuring data security (preservation of data from falsification, destruction and unauthorized access) and data protection (avoidance of unauthorized use of personnel-related data, see section 7.3.1.1)

- *Reliability*: The DBS has to be protected against system crashes. After a crash it has to be capable of restoring data consistency

## 3.1.7  Data Structuring

Due to the high costs of later changes within an already implemented database, planning and structuring of the technical requirements placed on a database are indispensable. Specifically, we refer here to the creation of a data model [Picot/Maier 94, pp. 115 et sqq.], i.e. a description of the segment of reality that is to be represented in the database. For this we utilize the Entity Relationship Model (ERM) described in section 6.4.1.2 in accordance with P. P. Chen [Chen 76]. It is ideally suited due to its low complexity and simple method to support the coordination process between central administration and different departments during the design of a data model.

In order to structure data using the ERM, data elements (entities) need to be identified and described based on their relevant characteristics (*attributes*), and their *relations*. Entities are individual or identifiable examples of things, persons or terms of the real or imagined world. Such an entity might be, e.g., the supplier "Smith" or a certain item. Entities with similar characteristics may be grouped into so-called entity types (e.g., all suppliers). The characteristics of individual entities or of an entity type are characterized with attributes. Relations may exist between or among entities (supplier Y delivers item 5).

The ERM is well suited to undertake initial structuring of the data (data modeling), but for this model there are no possibilities to describe data manipulations and data queries. Thus, using the ERM to develop a data model implies that this is only a preliminary stage of a database model.

## 3.1.8  Database Models

If the conceptual schema was modeled, subsequently it will have to be transferred into a database model. If such modeling occurs using the ERM we may utilize partially automated processes, available within most database models that take care of these transformations.

Currently available DBS are mainly based on the relational or the object-oriented database models.

### 3.1.8.1    Relational Database Models

The relational database model according to E. F. Codd [Codd 70] is based on the *Relations Theory* and thus on precisely specified mathematical principles. The columns of a table are labeled tuples. Within an ERM, a tuple corresponds to an entity. Each tuple must have a key by which it can be identified (primary key). Attributes of a relation are represented in the columns and a value range is assigned. Figure 3.1.8.1/1 depicts an example relation 'item' with the attributes ITEM_NUMBER (underlined as the primary key), ITEM_NAME, CLASS_OF_GOODS, and ITEM_PRICE.

From the definition of a relation, we may derive a number of characteristics:

- There are no two tuples in a relation that are identical to each other, i.e. the rows of a table are pairwise different.

- The tuples of a relation are not subject to any order, i.e. the sequence of rows is irrelevant.

- The attributes of a relation do not follow any order, i.e. the exchange of columns does not change the relations.

- The attribute values of relations are atomistic, i.e. they consist of a one-element quantity.

- The columns of a table are homogenous, i.e. all values within a column are of the same data type.

The transformation of a data model as an ERM into a relational model addressed above may be done following the rules as specified below [Ullman 95] (A detailed example is offered in the introduction on the Internet in this book. For a comprehensive example of these discussed relations see section 6.4.1.2.):

- In the relational model, an ERM-entity type is substituted through a relation whose attributes are identical with those of the entity type. If the entity type does not have its own key then special rules must be followed [Ullman 95].

- A relation in the ERM will be substituted in the relational model by a relation whose attributes consist of the keys (and not of all attributes) of the participating relations in ERM (cf., figure 3.1.8.1/1).

Relation "Customer"

| Customer# | Customer | Customer_city |
|-----------|----------|---------------|
| 081125 | Kaiser, K. | Boston |
| 732592 | Rainer, P. | Chicago |
| 002735 | Koenig, G. | Dallas |
| 773588 | Baron, R. | Rochester |
| 345764 | Graf, S. | Berlin |

Relation "Item"

| Item_Number | Item_NAME | Class of goods | Item_Price |
|-------------|-----------|----------------|------------|
| 15003 | QE 1300 | A | 598,00 |
| 37111 | CDP 100 A | B | 898,60 |
| 34590 | Sound 7 | C | 193,70 |
| 23676 | QE 1700 | A | 715,50 |
| 40400 | Quattro B | D | 5100,00 |

Relation "Orders"

| Customer# | Item_# |
|-----------|--------|
| 081125 | 40400 |
| 732592 | 23676 |
| 002735 | 34590 |
| 345764 | 37111 |
| 773588 | 15003 |

*Fig. 3.1.8.1/1    Description for a N:M Relation between the Relation "Customer" and the Relation "Item"*

- For a relation that was created following the first rule the key of the relation of the ERM is taken over.

- For a relation that was created from a 1:1 relation one may choose a key of the relations participating within the ERM.

- For a relation that consists of a 1:N relation one usually chooses the key of the second relation in the ERM.

- For a relation that consists of a N:M relation one chooses the combination of the keys of the participating relations.

DBS, based on the relational model (relational database systems), are characterized, when compared to older database models such as the hierarchical database model, by their high level of effectiveness and flexibility. They make an *easy variation of the relations's schemes possible*. Attributes may be added, changed or deleted. Relational models permit *manifold and simply executable data manipulations*. Thereby, users with little database knowledge may make queries and evaluations in a relational DBS.

Codd laid a cornerstone with the rules for normalization of relations for the teaching of data structuring within relational databases [Picot/Reichwald 91]. The goal of normalization is to structure a database such that the processing of data is simplified and that undesirable dependencies among the attributes of entity types during adding, deleting and changing of data do not occur. If such dependencies do exist and if they are not being considered during data manipulation (while, e.g., only a portion of the redundantly available data are deleted or changed) then bothersome data inconsistencies may result.

### 3.1.8.2    Object-oriented Database Models

A relational database model has a particular weakness: How may one store a complex entity such as an airplane, car or a house in a database? The cause of this weakness may be found in the saving format. Within a relational model everything is stored as a table. A car with all of its own characteristics and behaviors, however, can hardly be forced into a table. This is still possible for characteristics (such as color, number of wheels, etc.) as long as for each characteristic a column exists within the table; but how would one represent behavioral patterns of a car (such as braking, accelerating, blinking, etc.)?

In object-oriented database models "things" are stored (made persistent) as objects. Such object-oriented databases, having to meet the same requirements as relational databases, exist since the 1980's. In the beginning of the 1990's some software designers got together and formed the *Object Database Management Group* (ODMG-93) [Cattell/Atwood 94]. The then created standard still essentially determines object-oriented databases available today.

Objects in the object-oriented database model correspond largely to those described in section 6.4.1.4. Each object

- has at each point in time a behavior that is determined by operations that may be carried out on that object

- has at each point in time a condition, determined by the values of the object's attributes

- may be participating in relations with other objects

In contrast to the relational model there is also a relation component of the object itself (cf., fig. 3.1.8.2.1).

In the ODMG model each object has its own explicit type, just as in the relational model each tuple belongs to an explicit relation. The types are described via classes. These classes are defined using the *Object Definition Language* (ODL). Thus, the classes correspond with the schemas in relational models. A class, accordingly, determines the quantity of the potentially possible objects of a type.

1:N relations (see section 6.4.1.2) are supported by so-called aggregates. Since relations in the ODMG model are always bilateral, we also may implement N:M relations (see section 6.4.1.2) via two such aggregates in participating objects.

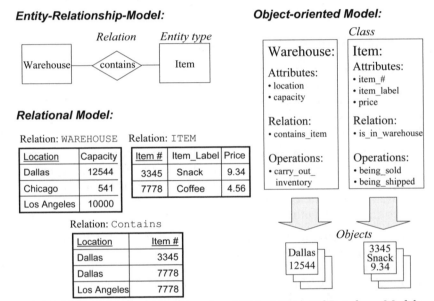

*Fig. 3.1.8.2/1    Relations in Relational and Object-Oriented Database Models*

Objects in the ODMG model have an identity that is dependent on their respective condition. This is realized via the attribute *oid* (object identifier). Different objects have different oid values.

To store, reclaim, and modify persistent objects there is the *Object Manipulation Language* (OML). Here we encounter for each object-oriented language its very own connection to the model.

Object-oriented databases have additional major advantages over relational databases: If something is being manipulated in object A and if this has effects on objects that have relations with object A, then these effects are executed automatically by the system. Moreover, although each object may be participating in several relations, it is represented only once in the database. In this way data redundancies may be avoided that may, otherwise, be the cause of considerable problems in relational data bases. Beyond that, object-oriented database models have the same advantages as described for object-oriented modeling (see section 6.4.1.4) and for object-oriented programming (see section 2.2.1.2). We want to emphasize in particular here the better reusability and the higher clarity.

Based on these advantages and the fact that complex structures, occurring ever-increasingly (multi-media databases, CAD see section 5.1.1.1), may be better represented in object-oriented database models and it follows that object-oriented databases are gaining increasing importance.

Also, so-called object-relational databases have just recently been put on the market. These are essentially relational databases that have been expanded using object-oriented components.

### 3.1.9  Query Possibilities for Database Systems

Depending on the type of the underlying database, differing query methods can help the user to access stored data. These methods are especially of importance if the user is expected to communicate directly with the database management system. The latter would be the case if no application programs were available that could undertake the data transfer on their own.

- The probably best known and widest distributed declarative (see section 2.2.1.2) query method for relational DBS is the *Structured Query Language* (SQL). It has all necessary structural elements to enable the user to choose among table columns and rows, as well as the linking of several tables among each other.

- Fundamental programming knowledge is required for the use of this query method, no matter if interactive in one session with the database management system itself or embedded into an application program (*Embedded SQL*). This is why SQL, in spite of all flexibility available to the user, is not suited for direct deployment at the workplace.

- Another possibility of relational database research is *Query by Example* (QBE). Here the user is given a sample table in which certain fields are filled out, others are left empty. The system then considers in searches the field contents provided by the user and inserts into the empty field suitable content from the database.

- In the area of object-oriented databases usually the communication with an application program is emphasized. These databases were primarily developed to store the administered objects beyond the point in time when the program finishes. In doing so, little attention was paid to the introduction and standardization of an interactive and effective query language. An exception is the *Object Query Language* (OQL) that, analogous to SQL, permits the selection of object quantities using arbitrary criteria that still need to be specified.

Example: We want to search for the names of all students who are enrolled in their sixth semester:

SQL:                                    OQL:

```
select x.name                    select x.name
from students x                  from x in student body
where x.semester = 6;            where x.semester = 6
```

The syntax of an SQL or OQL query is quite similar, although the two database models utilized have nothing in common. This becomes even more evident during more complex queries.

## 3.2 Networked Databases

As an essential characteristic of data integration we stated in section 3.1 that data should be organized only once. First we considered a central data organization. But when firms are distributed via multiple locations, a central data organization does not make sense, as data should—whenever possible—be managed in that location in which they are needed. The decentralized data organization lowers the cost for data transport and reduces the response times of database queries. Section 3.2.1 explains the peculiarities of *distributed databases*.

The *data warehouse* (see section 3.2.2) captures data from various, networked databases and offers tools to evaluate these for management decision making. Often users apply the *Online Analytical Processing* (see section 3.2.3) with the data warehousing software.

A discussion about the information generation from data that are available in a networked, but not integrated form, completes this section (see section 3.2.4 to 3.2.6).

### 3.2.1 Distributed Database Systems

Originally the Three-Levels-Architecture was developed for databases whose data were kept centrally in one location (see section 3.1.6). Using a minor variation it is also applicable with databases with data running on networked computer nodes communicable with each other in different locations, the so-called distributed DBS (cf., fig. 3.2.1/1).

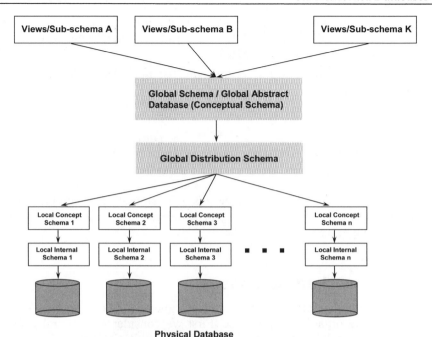

**Physical Database**

*Fig. 3.2.1/1      Three Level Architecture of Distributed Database Systems*

In contrast to the database system based on ANSI/SPARC the data are distributed with a distributed database system over multiple locations. This implies that to each location a *local internal schema* and a *local conceptual schema* are assigned. The *global conceptual schema*, known to each location, is dissected via a *global distribution schema* into the local schemata.

While the internal level and the external level are barely different from the ANSI/SPARC model, the conceptual level here has the additional task to dissect the local conceptual schemata.

Going beyond the requirements imposed on a database system in accordance with ANSI/SPARC (see section 3.1.6), a distributed database system based on C. J. Date [Date 95, pp. 598 et sqq.] is to fulfill the following conditions:

■  *Local autonomy*

Each computer is expected to have a maximum of control over the data in its memory/storage. Access to data is not supposed to be dependent on other systems.

■  *No centralized nodes*

In order to avoid bottlenecks central system functions have to be avoided. All nodes have to have equal permission and priorities.

■ *Failure-free operations*

Database operation is supposed to continue even if a node is completely disconnected. Moreover, the capability should exist to carry out configuration changes during database operations or other software updating.

■ *Location independence* (also: Location transparency)

The user is not expected to be informed at which location his/her data are physically stored.

■ *Fragmentation independence*

It should be possible to deposit different data records in different locations. In this way data may be stored in those locations in which they have their highest demand.

■ *Replication independence*

The user should not see if his/her data are present in multiple (replicated) copies in the database. This means that it is the task of the DBMS to make sure possible data is changed repeatedly during data changes such that inconsistencies will not occur.

■ *Distributed query*

Within the same query it should be possible to place database queries with multiple nodes.

■ *Hardware, management system and network independence*

The database system is to be independent of the utilized hardware, the management system and the network.

■ *DBMS independence*

The same DBMS does not have to run on all nodes. It is sufficient that all used DBMS have the same interface.

Distributed DBS can be used wherever data are captured decentrally and where the database queries concern the data inventory of other parts of the enterprise. For instance, this is the case in the automobile industry.

## 3.2.2  Data Warehouse

The focal point of today's DBS is the automation and the control of functions and processes (horizontal integration of the value added, cf., Fig. 4.2/1). Management, however, also needs support for decision making through specific information (vertical control of the value added) .

For the support of automation and the control of functions and processes usually only current data from a small segment of the firm is required (*operative* data). For decision making in the firm data are needed covering a larger enterprise segment, and they tend to cover a longer period of time. This task is taken on by the data warehouse within which the data are collected and prepared.

A data warehouse may be defined as a collection of

- subject-oriented
- integrated
- time-dependent
- non-volatile

data which may yield information for management decisions. This is why the data warehouse is also often referred to as the *information warehouse*.

In a bank one is likely to model such terms as CREDITCARD, BORROWER or SAVINGS. A data warehouse on the other hand is directed toward main subjects. With the above bank example the subjects might be CUSTOMER, SELLER or PRODUCT. With regard to the data or more specifically their relations to each other, there is one additional major difference to operational databases. Here we specify during the design phase which data are linked to each other. In contrast, one task of the data warehouse is to discover these relationships and linkages in the first place.

The very large quantities of data encountered in operational databases are often distributed in differing formats over a multitude of isolated information systems. Moreover, often the keys of the data records in the various locations of the enterprise are in disagreement with each other. Accordingly, it is possible that the very same customer is listed under several customer IDs. The data must be transformed appropriately such that they are available in their *integrated* form.

Data in a data warehouse are largely a collection of momentary snapshots of operational databases. Similar circumstances are documented with different data records. This is why the keys should always possess a time component, i.e. they are *time-dependent*.

An additional characteristic of a data warehouse, *non-volatility*, refers to the degree to which data may be updated. While in operational databases data may be changed or substituted at any time, data that once have been entered into the data warehouse will at most be deleted, but otherwise they will not be changed.

Data management deals with data provision and the transformation of data. Data of operational databases are adopted in cyclical intervals in a prepared form (selected and transformed) into the data inventory of the data warehouse. Besides such *internal* data there are also usually *external* data that may come from market research, for example. In newer systems one may also store qualitative information, such as verbal comments, that describe a remarkable time series.

A data warehouse consists of three fundamental elements: data management, data organization and evaluation/preparation (cf., fig. 3.2.2/1).

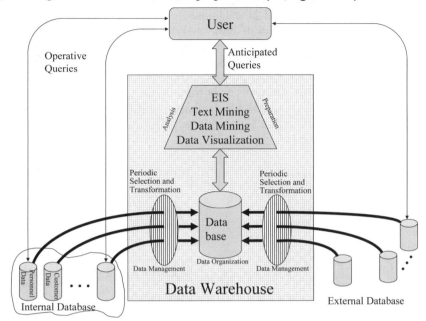

Fig. 3.2.2/1    *Architecture of a Data Warehouse*

The data organization describes how to treat the data physically and logically. Questions may be clarified pertaining to access and forms of storage. Accordingly, data organization corresponds with the lowest level of the Three-Levels-Architecture of DBS (see section 3.1.6). It should also be noted that due to the sizable data quantity other problems may appear here, such as the design of the physical database.

The third fundamental element of the data warehouse is the evaluation/preparation. This element poses a special challenge, for queries are to be formulated as much as possible in natural language in order to minimize costly training and familiarization for managers. This means, however, that queries will have to be semantically evaluated (i.e. information needs to be generated from data). This is not the case with operational DBS, since the semantics there are induced via the syntax of the query language.

The most important tool of evaluation/preparation is *data mining*. Data mining is the analysis of the data inventory in order to discover relationships among the data that have not yet been discovered. Data mining is a very complex process, which may also be used for text (*text mining*).

### 3.2.3  Online Analytical Processing

An important technology for the data warehouse and for management support systems (see section 4.3.2.2) is *Online Analytical Processing* (OLAP). With many other modern forms of data analysis we encounter the problem that they need many time-consuming, sorting, selection and compression processes in order to be able to analyze multidimensional data (e.g., sales of product groups per customer group). If these processes are not conducted until the actual analysis run time, then the response time of these processes increases drastically. In an OLAP database, with the appropriate settings, all conceivable condensings efforts are ready for analysis as cells of a multidimensional data cube (cf., fig. 3.2.3/1).

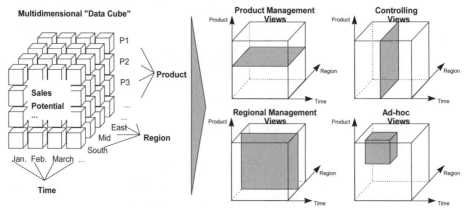

Fig. 3.2.3/1    *Different Views on OLAP Cubes through multidimensional Analyses*

At the same time OLAP's multidimensional view reflects the perspective of a multidimensional business management analysis quite appropriately. All possibilities of viewing and examining the data are already manifested in the pre-calculated values. Each other slice or dice configuration represent then different questions. This is why OLAP may be used as a building block within the evaluation/preparation in the data warehouse (see section 3.2.2).

### 3.2.4  External Databases and Information Retrieval

Frequently one requires information for entrepreneurial decisions that are only externally available (e.g., about competitors, markets, scientific, legal, technical or political developments). Suitable information sources for this are *external databases* which are usually used in interactive dialogs [Mertens/Griese 00, pp. 27-33 et sqq.]. They have found widespread use in the areas of science, economics, technology, law and patent concerns. U.S. database vendors dominate European information services in this area. *Ac-*

*cess* to stored data inventories is associated with costs and requires suitably equipped data transfer in international computer networks (see section 2.4).

It is important to distinguish between *full text* and *reference databases*. While in full text databases the documents in their entirety (data) will be available, the latter merely offers biographical hints, key words or headline text and, occasionally, short summaries of the searched text.

Since external databases offer their data inventories usually in *heterogeneous form*, i.e. not in a particular structure (see section 3.1.6), a special technique for the recovery of information is required by the database user. An *information retrieval system* supports in particular the physical organization of unformatted data.

The search by users for particular documents is based on defined *descriptors* (a designation for the description of texts; e.g., author names or headline words) and their logical connection (e.g., via "AND" or "OR", respectively). Example: "Find all articles by the author Smith about cost accounting in the pharmaceutical industry in English"; for this the descriptors "Smith", "cost accounting", Pharmaceutical Industry" and "English" must be linked by AND. If articles aside from the pharmaceutical industry are also desired from the "chemical industry", the descriptors "pharmaceutical industry" and "chemical industry" need to be linked by an OR. If a search yields too many results, then instead of the too generally formulated descriptors (e.g., "cost accounting") more specified descriptive words (e.g., "direct cost accounting") or additional, limiting descriptors (something like "since 1990") must be used.

## 3.2.5  The Internet as a Database

Databases may be classified as internal and external databases (see section 3.2.2). After more careful examination of external databases (see section 3.2.4) we may discover that it will be difficult to recognize a clear-cut operational or a decision-support character. For example, it would be possible to view the homepage of an academic department in the broadest sense as an external database (e.g., currently offered courses can give information about the current research emphasis in the department). A homepage, however, also contains decision-supporting data (e.g., data about past courses), as well as operational data (e.g., data that may be incorporated dynamically from the faculty database into the homepage).

But also the classification along external and internal databases is not always clear-cut and has a subjective component. If, for example, two firms work together in the area of supply chain management (see section 5.4) within an Extranet, then one firm will make available a good portion of its operational (internal) database, but for the second firm this database will become an external database.

Still more complex is the situation when viewing all databases collectively. These are in some form interconnected via the Internet. In this way one may view the Internet as an enormous "distributed database". It must be marked however that this database only partially fulfills the criteria of a distributed database. The Internet is in the strictest sense not consistent as a distributed database, but rather is "anarchically" organized. Therefore there are no global conceptual schema and also no sub-schema. Nevertheless certain services and tools permit us to utilize in a limited fashion the Internet like a database during the search for data and information (see section 3.2.6).

## 3.2.6 Research on the Internet

*Search engines* permit the targeted searching for information on the Internet through the use of key words. Prominent examples are AltaVista™, Google™, Lycos™ and Yahoo™. They may be called up directly just like other information pages. Some web browsers (e.g., Netscape™ Navigator, Opera™) offer integrated forms to the most important search engines.

The technical implementation of search engines varies. Generally descriptor information for the search tools are read off the available web pages in the Internet actually *before* the search is initiated by the user. The search engine administers its own key word database that is constantly updated in the background by searching the Internet. The creator of the web page specifies his/her own key words. The language HTML (see section 2.5.2) contains a construct with which the descriptor information for search engines may be made available. In addition, the search engine also indexes the page content. Alternatively, the possibility exists to directly inform the vendor of the search engine of the existence of one's own pages.

Meta search engines take an extra step in that they utilize several search engines simultaneously. An example of this is "MetaCrawler".

A big problem for vendors of search engines is the preparation of a hit, i.e. when a match based on the key word(s) is found. Since the Internet has today a no longer manageable number of websites, hits are usually quite numerous. But then, which hits are for the user truly relevant? The ranking of hits is an immensely complex task such that many vendors take other approaches and select the found information a priori. This is partially accomplished by setting search engines aside for specific theme areas. An important example is CNet™ that offers a search engine for the area of information technology on its portal.

The use of XML (see section 2.5.2) makes better evaluation of hits possible. The description of XML may contain content, i.e. semantic information besides the presentational forms common in HTML. Consequently, search engines should have an easier job preparing information. If a search contains the word *Golf* and it can be determined that this relates only to automobiles, search engines could then sort out hits referring, for example, to the sport of *Golf.*

An additional problem for search engines is the trend toward dynamic web pages. While with static pages information is fixed in HTML or XML with dynamic pages the activation of the page triggers a webserver to extract certain information from a database. This offers the advantage that it doesn't lead to data redundancy, and in this way the corresponding web pages cannot become inconsistent. The connection to the database occurs then via an appropriate interface. The technology for the translation of interactive web pages is presently being taken care of by varying processes.

*Active Server Pages* (ASP*)* by Microsoft offers so far the most widely used application variant. Although ASP may only be deployed in unrestricted functionality, Microsoft's webserver and the SQL database server may be used. Considerably more liberal is the oldest of the three current technologies, the *Common Gateway Interface* (CGI) standard (see section 2.5.2). This procedure is though the most rudimentary of the three technologies which also explains why it is more and more eased out of the market. *PHP* (recursive acronym for *PHP: Hypertext Preprocessor,* formerly *Personal Home Pages*) is within the Open Source area the most important alternative to ASP. In contrast to the CGI standard which is not based on its own language (it is possible to program with e.g., Perl), PHP has its own specifically developed and easily learnable programming language.

If dynamic web pages are also interactive, i.e. the user specifies which information is desired, then search engines will reach their limits because they do not support such an interactive search. Since usually very large databases are being made available via these interactive, dynamic pages, one estimates that the overwhelming portion of the available data in the Internet is not indexed via the most common search engines and thus also cannot be traced. At the present time this is a major challenge for the designers of search engines.

The technology of search engines may also be deployed internally within a firm in an *Intranet* to fulfill information needs. Appropriate software is designed similarly to that of the large Internet search engines.

Search engines have also been recently integrated with *portals.* A portal is a website on the Internet in which a wide spectrum of services, aids and tools are made available. Aside from search engines, this may be, for example, e-mail, discussion forums, stock exchange and weather information, and also places to play games. The goal of portals is to serve as entry points for users to the Internet, but may become the sticking points. In this way a number of *hits* are generated that may serve as a means for advertising revenue.

## 3.3 Literature for Chapter 3

Cattell/Atwood 94          Cattell, R.G., Atwood, T., The Object Database Standard
                           ODMG-93, San Mateo 1994.

Chen 76                    Chen, P.P., The Entity-Relationship Model: Towards a Uni-
                           fied View of Data, ACM Transactions on Database-Systems
                           1 (1976) 1, pp. 9-36.

Codd 70                    Codd, E.F., A Relational Model for Large Shared Data
                           Banks, Communications of the ACM 13 (1970), pp. 377-
                           387.

Date 95                    Date, C.J., An Introduction to Database Systems, 6th edi-
                           tion, Reading/MA 1995.

Mertens/Griese 00          Mertens, P., Griese, J., Integrierte Informationsverarbeitung
                           2, Planungs- und Kontrollsysteme in der Industrie, 8th editi-
                           on, Wiesbaden, Germany, 2000.

Picot/Maier 94             Picot, A., Maier, M., Ansätze der Informationsmodellierung
                           und ihre betriebswirtschaftliche Bedeutung, Zeitschrift für
                           betriebswirtschaftliche Forschung (ZfbF) 46 (1994) 2, pp.
                           107-126, Verlagsgruppe Handelsblatt, Duesseldorf, Germa-
                           ny.

Picot/Reichwald 91         Picot, A., Reichwald, R., Informationswirtschaft, in: Heinen,
                           E. (Hrsg.), Industriebetriebslehre, 9th edition, Wiesbaden,
                           Germany 1991.

Ricardo 90                 Ricardo, C., Database Systems: Principles, Design and Im-
                           plementation, New York 1990.

Scheer 94                  Scheer, A.-W., Business Process Engineering – Reference
                           Models for Industrial Companies, Berlin, Germany, 1994.

Ullman 95                  Ullman, J. D., Principles of Database and Knowledge-Base
                           Systems, Volume I, 8th edition, Rockville 1995.

Wigand et. al. 97          Wigand, R.T., Picot, A., Reichwald, R., Information, Or-
                           ganization and Management: Expanding Markets and Cor-
                           porate Boundaries, Chichester, et al. 1997.

# 4 Goals, Forms and Means of Integrated Data Processing

## 4.1 Goals

The word "*integration*" means "reestablishment of a whole". In the context of business information systems integration is to be understood as the connection of people, tasks and technology to a consistent whole [Heilmann 89]. From an enterprise perspective, functional, process and departmental borders and boundaries are more or less artificially determined, i.e. utilitarian principles will determine such borders and boundaries. Integration efforts enable the diminishing of negative effects of these borders and boundaries. The information flow may be viewed as a natural reflection of all processes in the enterprise that actually belong together. For example, in order to take the opportunity to reduce the division of labor in a firm, several tasks are combined in one place.

A special advantage of integration efforts is the opportunity to reduce the costs of manual data entry and to keep them at a minimum. This is because within an integrated conceptual framework the single software tools deliver the largest part of the data in machine-readable form. Accordingly, the accounting system receives wage data from the wages payroll, the evaluated material flows from the material valuation, accounts receivable from the invoicing, accounts payable from the supplier invoice control, etc. (cf., also fig. 5.1.3.1/2). Since each transaction automatically triggers a subsequent transaction nothing will be "forgotten". For example, it cannot happen that after a customer received a charge-back the corresponding correction of sales data is forgotten. Those advantages of integration may also be seen as troublesome in that incorrect data inputs will reproduce themselves into many programs like a chain reaction.

## 4.2 Forms

The integration described in section 4.1 is called *functional integration*. Occasionally the terms *procedure integration* or *process integration* are used as well. Chains of procedures or *business processes* in the firm or sometimes also transactions among companies are triggered automatically to a large extent. In the production firm this particularly impacts the four so-called main or key processes (cf., fig. 4.2/1):

■  Offer process (customer-to-order)

■  Order process (order-to-invoice, cf., fig. 5.1.5.1/1)

■  Product development (idea-to-market, cf., fig. 5.1.5.1/1)

■  Customer service (failure-to-invoice)

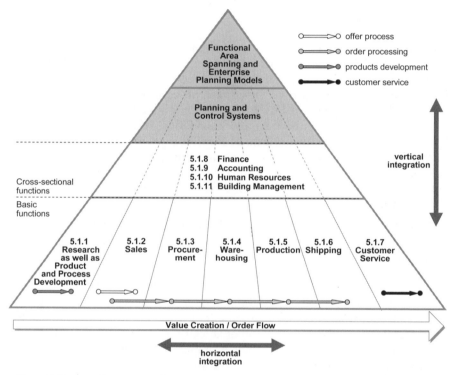

*Fig. 4.2/1*      *Direction of Integration*

In chapter 5 we will get to know these forms of integration through numerous examples from different industries. Functions which are represented on the computer with the aid of software programs operate on data stored in memory. Thus *data integration*, already mentioned in chapter three, is the counterpart to function and/or process integration.

Additional developments of integrated data processing can be defined. They partially reflect less the views of the business manager but rather those of the information systems specialist (for additional forms of integration and more details see [Mertens 01]):

1.   The *integration of different information representations and media* uses the newer technical possibilities in order to represent, store and

transfer text, diagrams, audio and video sequences together. For example, for product presentations a computer is used showing a film over the article on its screen, fading in technical data in tabular form and playing music in addition. This technique is called *multimedia*.

2.     The *integration direction* in the pyramid (cf., fig. 4.2/1) shows the organizational architecture in the example of a manufacturing firm. We may differentiate here between horizontal and vertical integration. The numbers in the figure refer to the sections in this book in which these topics are described.

      2.1     *Horizontal integration* is primarily the connection of the administrative and the disposition systems (see chapter 1) in the production value chain. In the manufacturing firm business processes are mainly to be integrated in the transactions originating from customer orders. They begin with the offer process and end with the posting of the customer payment.

      2.2     *Vertical integration* particularly refers to the data supply of planning and control systems from the administrative and the disposition systems.

3.     Related to the *scope of integration* we need to differentiate between the internal and the interorganizational (cf., fig. 4.2/2) integration (see chapter 1).

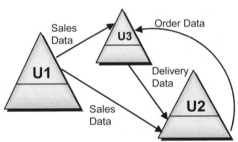

*Fig. 4.2/2*     *Interorganizational Integration*

For example, the computer of a producer of car spare parts $U_2$ uses sales data from a car manufacturer $U_1$ for its warehousing operations. Then it passes the result on to the computer of subcontractor $U_3$, from which the spare part producer orders steel sheets (see *Supply Chain Management* (SCM) in section 5.4). $U_3$ receives from $U_1$ simultaneously sales data for his long-term planning and informs the $U_2$ warehouse which deliveries are forthcoming. Those kinds of networks can be found especially in virtual companies (see section 7.1.1.1).

4.     Based on the *degree of automation* we may differentiate between fully automated and partially automated information transfer.

4.1     *Fully automated* information transfer, e.g., exists if a user monitoring program (see section 5.1.5.11) triggers another software program when significant target deviations are present.

This program then provides a "diagnosis" and starts a suitable "treatment", i.e. a rescheduling measure, to remedy the situation.

4.2     In the case of *partially automated* solutions humans and machine cooperate. Again we need to differentiate *who* subsequently triggers an action. Usually a planning executive takes the initiative. This may be the case, e.g., if he or she recognizes an initiating delay in the purchase process and reacts to this with a reminder sent to the supplier. With *workflow management systems* (see section 4.3.1.2) personnel-based actions are triggered by the information system.

# 4.3    Methodological Aids

## 4.3.1  Systems for Procedural Support

### 4.3.1.1    Transaction Systems

The purpose of transaction systems is the handling of formalized and short processes in a dialogue. These processes are often repeated, whereas only the input data frequently change [Meyer-Wegener 88, p.14]. Examples are the booking of a flight, the ordering of an article, the transferring of an amount of money or the documentation of a medical standard examination. During a transaction a preplanned dialogue is completed for the implementation of a business process. A transaction may consist of only one but also of many dialogue steps (involving persons or programs).

### 4.3.1.2    Workflow Management Systems

Workflow management systems (WMS) support the division of labor within well-structured business processes.

First of all special tools help to describe business processes: For example, the workflow in the preparation and delivery of an offer. The procedures are then completed via partially sequential and partially parallel execution of the modeled steps. The involved employees automatically receive the necessary information. The associated support systems for such regulated division of labor processes carry out the following important functions:

- *Generation of procedures*: Selection and release of a suitable type of procedure. For example, if a customer inquiry arrives, the procedure "inquiry-processing" is started.

- *Procedure organization and control*: Dismantling of the procedure into an action network, triggering the respective procedural steps to be enacted and possible procurement of the necessary information (*information logistics*). The routing decision (where does the information travel to next?) can be chosen by people or be triggered by programmed rules within the system. Two examples describe this:

  - A credit clerk of a bank feels uncertain, because a high loan amount is to be granted. Therefore he/she decides to place the procedure into the electronic mail box of his/her manager.
  - It is specified in the WMS that all credit requests exceeding $100,000 are automatically conveyed to the manager.

- *Procedure information and tracking of follow-up*: Making information available about handling a process, as well as due date monitoring. For example, an executive has access to a graphic overview of the current processes on his/her computer screen.

- *Procedure closure*: Completion of the procedure and possibly the consolidation of partial results to total results.

### 4.3.1.3 Document Management Systems

Document management systems (DMS) are designed to save documents in electronic form, and to administer and to locate them in the computer's memory with the help of descriptors (see section 3.2.4). Documents, which are still in paper form, are frequently captured electronically through the use of scanners (*imaging*).

Optical memory disks are a suitable storage medium and may be unified in so-called jukeboxes. Different electronic documents, e.g., customer inquiries and corresponding CAD-designs (see section 5.1.1.1) are unified within an electronic procedure folder.

### 4.3.1.4 Workgroup Support Systems

Here we are concerned about computer-based methods in the support of teams during processing of a common, relatively unstructured task.

Cooperation is based on network architectures with associated communication systems. Important support systems for workgroup tasks are:

- *Conference planning systems*: Support of team members with a calendaring function, e.g., with common appointment calendars, as well as management of necessary resources such as rooms, presentation or communications devices.

- *Computer conference systems*: Asynchronous (e.g., e-mail) or synchronous (e.g., videoconferencing) discussion between or among spatially separated persons.

- *Group Decision Support Systems* (GDSS): Tools for decision making by teams.

- *Co-Authoring*: Tools for the common handling of, e.g., texts, plans or graphics. Changes are immediately communicated to the team members and several team members are able to work on different sections of the document at the same time.

In addition, numerous tools are available, including, e.g., systems for meeting moderation, common file systems, project management software and different variants of e-mail.

### 4.3.1.5   Knowledge Management Systems

Knowledge management systems are used in order to make existing knowledge from the enterprise available in a convenient form. They integrate knowledge, which is usually dispersed widely throughout the enterprise (e.g., in databases, documents, process documentations and in the heads of co-workers). Thereby they particularly utilize "knowledge about the knowledge" (meta-knowledge) as well. For that purpose functions and processes are specified and supported for development, representation, administration, transformation and refinement of knowledge. Basis of a knowledge management system is an "enterprise memory," that independent of applications stores information and knowledge of the entire enterprise, as, e.g., in a data warehouse (see section 3.2.2). Thereby the storage covers different representational forms, as well as the pertinent meta information.

Content management systems are an example of knowledge management systems. They permit, e.g., a distributed administration of Intranet, Extranet and Internet sites. For that purpose they offer the following functions: checking of consistency of links, separation from content and layout, integrated user administration, programming interface for individual extensions (e.g., for connection of databases), administration of websites with a browser, as well as a "staging concept". This means that new versions for websites are developed parallely and tested on the up-to-date version and published only after successful conclusion of tests.

### 4.3.2  Planning and Control Systems

In comprehensive and mature integrated systems especially administrative systems store numerous data from all operational ranges in databases or data warehouses (see section 3.2.2). Such data may be used in the sense of a vertical integration for the information of managers and for providing data for programs which, in turn, may suggest to managers ideas and proposals for potential meaningful decisions.

However, modern planning and control systems are not only provided with internal, but also with external data (see section 4.3.2.1).

## 4.3.2.1    Types of Planning and Control Systems[1]

Figure 4.3.2.1/1 contains an attribute collection somewhat in accordance with a morphological box. In the following the box is only described in so far as the entries are not self-describing.

Planning and control systems may be triggered, if special data constellations arise. We also speak of *data or signal-driven* systems. The most important form are reports following the principle of "information by exception", i.e. deviations from plans, expected values, historic data and other comparative values that may trigger actions within the system.

Another common form of signal-driven reports are early-warning systems: Management specifies indicators and/or combinations of indicators that enable a timely warning notification under certain conditions by pointing toward important data patterns.

Systems that produce reports for certain scheduled *calendar dates* (e.g., at monthly or quarterly intervals) are probably the most frequent form of management information systems.

In many cases reporting is targeted to *individual persons*, in particular managers or executives or those who are assigned particular roles. The *user* is able to recall management information from the system, e.g., to start a control measure or when a decision is to be prepared.

With group-decision-support systems a *group* of users communicates with each other and with the computer (see section 4.3.1.4).

Usually the users of a planning and control system are middle management level employees. The need for information and methods for decision making at the *upper management level* is usually taken care of partially with underlying procedures of enterprise planning.

A number of estimations show that *external information* has an equivalent or even a higher value for managers than internal information [Bauer 96]. In conjunction with information processing, these concerns gain greater importance in that online databases and especially the Internet create possibilities to receive external information in machine-readable form.

---

[1] This section is an abridged and simplified chapter from [Mertens/Griese 02].

A challenge in planning and control systems is to combine *quantitative and qualitative* (e.g., press reports) information from internal and external sources.

| Trigger | Signals/Data constellations | Calendar dates | | User requests | | Decision needs |
|---|---|---|---|---|---|---|
| Number of Addressees | Individual | | | Groups | | |
| Addressee Hierarchy | Lower management levels | | Middle management levels | | Upper management levels | |
| Origin of Information | Internal sources | | | External sources | | |
| Information Type | Quantitative information | | | Qualitative information | | |
| Presentation Form | Messages | Spread-sheets | | Charts | Verbal reports | Evalua-tions |
| Query Mode | Standard queries | | | Open queries | | |
| Information Distribution | Pull method | | | Push method | | |
| Decision Model | Not present | Decision models using statistical methods | | Decision models using Operations Research methods | Decision models using AI methods | |
| Simulation | No simulation | | | What-if-calculations | How-to-achieve-calculations | |

*Fig. 4.3.2.1/1    Morphological Box (according to [Mertens/Griese 02])*

The dominating presentation form are *tables*. There is danger in that too little care is exerted and attention paid to the conceptualization of planning and control systems. As a result we may encounter the dreaded "number cemeteries". This may especially happen in systems that generate scheduled management information at certain calendar dates. With modern report generators information may be edited automatically from statistics in tables into *diagrams*. It is more demanding to derive *verbal reports* (text generation) from statistics with the aid of expert system technology (see section 4.3.2.3). *Expertise systems* then are systems through which analyzed data are compiled in the form of evaluations.

If a user queries a database, the query may be *standardized* and preprogrammed. This means that at least the type of query must be already known during system planning. Briefing books are reports targeted for high level managers that are based on standardized queries. Query systems with *free data inquiry* have the restrictions of standardization omitted. The user may express his/her information needs while indicating which characteristics (descriptors) the sought-after information is to combine in that specific search (see section 3.2.4).

With regard to information distribution we differentiate between *pull and push methods*. With pull methods (passive) the user needs to get the necessary information on his or her own. With push methods (active management information systems) the system defines when which specialists and managers are to be informed and then "push" this information directly to the user.

In simple management information systems only data are supplied. Many planning and control systems, admittedly, provide decision support systems as well (see also section 4.3.2.2 on DSS). Sometimes knowledge-based elements as a form of *artificial intelligence* are integrated into conventional planning and control systems. This promises efficiencies everywhere where the problems themselves are badly structured, but well definable.

In many decision situations it is advisable to analyze alternatives by using a simulation. With *what-if calculations* the impact of a quantifiable measure is to be estimated. For example: "With what percentage of profitability will a line of business grow, if the price of products can be increased by one percent without the corresponding sales volume decreasing?" *How-to-achieve calculations* allow us to analyze alternative measures with which these goals might be reached. For example: "On average around what percentage must the contribution margin of a certain product be increased in order to raise the line of business profitability of a product group by one percent?"

## 4.3.2.2    Aids for Preparing Management Information

### 4.3.2.2.1    *Determination and Representation of Report Objects*

In order to prevent that managers experience information overflow caused by the computerized system, the information must be filtered. An important technique, in terms of the above-mentioned *information by exception*, is to report deviations from usual forecast or planned results. The definition which deviation represents an exception may take place in two ways: On the one hand, *tolerance limits* may be established the exceeding of which leads to the fact that this deviation becomes an exception. For example, all sales of an item group that deviate more than 5% from the plan are issued. On the other hand, we may define the exception in a variable way. For instance, we may want to define the ten largest deviations as an exception. Thus *hit lists* are being created [Mertens/Griese 02, pp. 73-78 et sqq.].

Sometimes an information system is confronted with the problem that exceptions appear on higher *aggregated levels*, e.g., in the form of performance results at the country level. In this case it is the task of the information system to systematically analyze the specific aggregation-hierarchies from top to bottom, in order to detect the range where the deviation arises (e.g., main item group ⇨ item group ⇨ item; customer group ⇨customer; state ⇨ region ⇨ sales area ⇨customer ⇨item). This takes place with the aid of so-called *navigators* (*drill-down-technique*). In conventional management information systems the computer has an exactly specified search path. Here, however, the challenge for the computer is to detect notable data constellations autonomously. For instance, the bad operating result may be mainly traced back to the fact that item 6633 was inadequately sold to key accounts in the southern region. Methods that detect the coincidence of notable characteristics use statistical methods like *cluster analysis*. Such techniques are summarized under the label *data mining* or sometimes also *business intelligence*. Accordingly, one imagines that the researcher "digs in an enormous data mine". From a software perspective this data mine may be organized as a *data warehouse* (see section 3.2.2).

### 4.3.2.2.2    *Management Support Systems*

Under *management support systems* (MSS) we understand all deployment techniques within business information systems that support management tasks. In the following these are identified:

Decision support systems (DSS) are interactive systems that assist specialists and managers in badly structured decision situations.

Management information systems (MIS) may be seen as a sub-part of MSS and as a counterpart to DSS. They support managers with data and/or information (data support). Data sources are here, aside from external public or private databases and especially the Internet, the operational databases and data warehouses (see section 3.2.2) within the company.

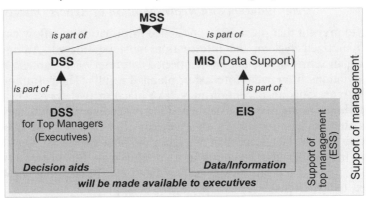

*Fig. 4.3.2.2.2/1 Systematic Depiction of Management Support*

A subset of MSS are executive support systems (ESS). They are especially developed for members of upper management. Executive information systems (EIS) have their main focus on data supply. They may be understood as

MIS equivalents, as far as top management is concerned. They usually have a uniform and user-friendly PC interface that integrates different forms of electronic reporting (see previous section as well as 5.1.13) with typical auxiliary functions, e.g., electronic mail.

### 4.3.2.3 Methods of Artificial Intelligence

Expert systems (XPS) or knowledge-based systems (KBS) belong to the field of artificial intelligence (AI). It is the goal of these systems to capture and store specific knowledge of human specialists in the *knowledge base* of a computer and use it for solving a multitude of problems. Usually the knowledge is stored in the form of if-then-relations, i.e. so-called production rules. A simple example from a offering system for cars is: "IF customer wants air-conditioning *and* electrical sunroof *and* electrical windows, THEN make an additional recommendation: consider buying a stronger battery or deselect one of the features chosen".

Essential is the separation of *knowledge base* and *problem solving component*. The problem solving component "traverses" the base of *domain-related knowledge* (this contains, e.g., the relationship between equipping the car with extras and the need for a stronger battery) under consideration of the *case-specific knowledge* (these are in our example the dialogue between the automobile salesperson and the expressed wishes and specifics of the customer). The problem solving component follows thus the often very complicated linkages of rules, until the system finds a satisfying suggestion or figures out that it may not find such a suggestion. The *explanation component* explicates to the user why the system arrived at a certain recommendation or decision (perhaps, in our example, because the standard battery may be unable to meet the heavy power demand during the winter months).

Imagine, you need a high-capacity computer but do not have the necessary money to buy a new one. Therefore, you may send a small search program out onto the Internet to find offers and auctions for used computers and to analyze the deals received. Maybe you even want to extend full authority to the program to make a buy decision when the PC it found is an especially good offer in meeting your requirements and price expectations.

Such a system, equipped with a certain degree of artificial intelligence, may be understood as a form of XPS and is also referred to as an intelligent agent. A central characteristic of an agent is its *autonomy*. That means the capacity to accomplish tasks, which were previously delegated to it, independently without the user having to pay additional attention. Other characteristics are the *communication capability* (agents exchange information among themselves and with human users), the *adaptability* to habits and work patterns of the user, the *mobility* (the above agent moves within the Internet) and in elegant versions a certain *learning aptitude*.

While the major task of our sample agent was searching and analyzing (see section 3.2.5), for other agents it may be part of their job to plan and control, perhaps within the scope of PPC-systems (see section 5.1.5.1).

Artificial neural networks (ANN), which often are briefly called "neural networks", try to duplicate stimulus-response and learning processes of the human nervous system. In doing so, special programs simulate a network comprised of switching elements which are modeled similar to human nerve cells. In a training phase users inform the ANN which output they desire with which corresponding input into the system.

For example: We enter the personality characteristics of job applicants that were hired in the past and had satisfactorily performed in certain workplaces. The system regulates its switching cells (so-called neurons) with the aid of complicated mathematical methods in such a way that certain characteristic combinations on the input side lead to a desired output (in this case: recommendation for a suitable workplace). When the training phase is finished the ANN may be used for decisions or recommendations in the real world. In our case jobs are suggested to applicants with certain characteristics.

### 4.3.2.4 Methods of Operation Research and Statistics/Method Databases

Many operation research methods are components of application systems. For instance, *linear programming* helps with the minimization of waste, e.g., in paper, glass or metal foil manufacturing. Methods of mathematical/statistical forecasting are useful for sale, outward stock movement and liquidity predictions. Particularly in the analysis of alternative possibilities in shop floor scheduling (see section 5.1.5.8) or delivery planning (see section 5.4) *simulation* as a method is winning increasing importance.

Operation research methods and statistical methods are not programmed for each application from scratch. They are stored as program modules in a *method database*. Just as a database is more than a collection of data (see chapter 3), a method database offers additional support for users as well. Included in such a database typically are systematic method directories, computer aids for the selection of procedures for a particular problem, the linkage of singular methods to larger models and explanations for using an algorithm, for setting a parameter, as well as for the evaluation of the results. An interesting example for a method database is the *Advanced Planner and Optimizer* (APO$^{TM}$) from SAP AG. This becomes important for supply chain management (see section 5.4) in which APO plays an important role.

## 4.4   Literature to Chapter 4

Bauer 96       Bauer, M., Altbekanntes in neuer Verpackung?, Business Computing o.Jg. (1996) 4, p. 46.

Heilmann 89       Heilmann, H., Integration: Ein zentraler Begriff der Wirtschaftsinformatik im Wandel der Zeit, Handbuch der modernen Datenverarbeitung 26 (1989) 150, pp. 46-58.

Krallmann et al. 97       Krallmann, H., Mertens P., Schiemann, I., Entscheidungsuntestuetzendes System (EUS), in: Mertens, P. (ed.), Lexikon der Wirtschaftsinformatik, 3rd ed., Berlin, Germany, 1997, pp.149-150.

Mertens 01       Mertens, P. Integrierte Informationsverarbeitung 1, Administrations- und Dispositionssysteme in der Industrie, 13th edition, Wiesbaden, Germany, 2001.

Mertens/Griese 02       Mertens, P., Griese, J., Integrierte Informationsverarbeitung 2, Planungs- und Kontrollsysteme in der Industrie, 9th edition, Wiesbaden, Germany, 2002.

Meyer-Wegener 88       Meyer-Wegener, K., Transaktionssysteme, Leitfaeden der angewandten Informatik, Stuttgart, Germany, 1988.

# 5  Integrated Application Systems

In the fifth chapter of this book we would like to give an impression of the many applications of information systems in various industries. In doing so we choose the most important application systems we are likely to encounter in the business world and try to represent their different types, i.e. administration, scheduling, planning and control systems, adequately.

We may take a look at the practice of business by making "slices" through the economy. In the field of statistics the terms "industrial production" and the "service sector" are being used. A differentiation that is not very selective and shows partial overlaps. Especially larger firms operate with elements of "both worlds". Surely it should be possible to cover various functions for both sectors of business at the same time, e.g., cost center accounting. Expensive marketing procedures, covered in section 5.2.3 for the service delivery process, are just as conceivable for the manufacturing firm and are indeed used there as well. In doing so we would, unfortunately, present the reader with a rather abstract description. Let us choose, e.g., product design: there is of course the design of insurance products, but this function or the process that has to be carried out is only comparable at a very high level of abstraction with computer assisted design (CAD) in the automobile manufacturing area.

Since there are no better delineations conceivable we accept here the partitioning into production businesses (section 5.1) and service businesses (section 5.2).

## 5.1  Application Systems in the Industry Sector

When dealing with application systems in production firms, first and foremost the value chain serves as an organizing principle, i.e. we shall follow basically a product from its development up to customer service in the after-sale phase and will show how information systems are characteristic for each respective operational area. Due to space limitations we cannot go into the particularities of different types of manufacturing firms (e.g., made-to-order or mass production) or the different sectors. Nevertheless, we choose our examples in a way that allows the reader to gain many varied impressions of the particularities of the sectors.

## 5.1.1  Research and Product Development

### 5.1.1.1  Product Design (CAD/CAE)

The focal point of product design is the *Computer Aided Design* (CAD).

CAD systems may be viewed as the transformation of the construction design from the drawing table onto the computer monitor ("intelligent drawing board"). Here the engineer may access all possibilities of modern computer graphics. The system, e.g., draws circles and other geometric formations after certain parameters have been entered (with the circle: coordinates of the center point and the radius; with straight lines: the coordinates of two points) on its own and shades marked areas at the push of a button among many other things. The results of CAD procedures are stored drawings or geometric data and the listings of the rendered designs (see section 5.1.5.3).

An important enhancement is *Computer Aided Engineering* (CAE). Here the designed product is represented as a model in the computer and it is thus possible to run simulations. CAE is used, e.g., to model the geometry of an automobile's body. The system then determines through the use of engineering calculations which effects a stronger slant of the windshield might have on air resistance and thus subsequently on maximum speed, gasoline consumption and the heating-up of the passenger cabin. CAE systems are complex information systems applications, as all interdependent product characteristics must be modeled within the computer. However, a well designed and developed CAE system leads to high-level efficiencies. In the just mentioned example the construction of car body variations is unnecessary that, otherwise, would have to be tested in the wind tunnel or even on the road (which is especially expensive).

The current trend is to develop CAD/CAE into comprehensive design information systems. Aside from the technical calculations we encounter so-called rapid or quick costing modules via which design and model variants may be examined for cost advantages.

An additional goal of such systems is to make sure that the total number of parts to be managed does not grow too quickly. One therefore supports the designer via internal and external databases (see section 3.2.4) with information about available parts, especially standardized parts, and about their utilization in other products.

In doing so an effort is being made that early-on during the design phase the engineer uses already existing design elements ("reusable components"). Additionally, the engineer may access external information from the Internet (see section 5.2.5), e.g., the behavior of material under large temperature fluctuations.

Under favorable conditions it is possible to let the computer construct or configure the product largely on its own; in modern solutions, e.g., with the assistance of XPS (see section 4.3.2.3).

*A PRACTICAL EXAMPLE*

*Kennametal Hertel AG manufactures tools, e.g., drills and thread milling machines. Among the company's customers are firms from the automobile, aircraft, machine manufacturing and heavy industries, among others. For special tool design and construction the expert system TESS (Tool Expert Software System) was created. The company's knowledge base includes experiences based on more than 100 person years of tool developers, more than 10,000 machine processing cases and more than 100,000 produced items.*

*Features desired by the customer of the tool, e.g., its shape or geometry, respectively, including measurement tolerances are inquired about by the system in a sequence of menus from the employee in charge of processing the order. In addition it examines in which tool machine the product is to be clamped into. The knowledge-based system produces, e.g., a drawing for the customer, a spreadsheet for making costing decisions, a rough work plan for sequencing tasks in the manufacturing process and the NC program for the controlling of the tool machines [Mertens 01, p. 34].*

The designer in the mechanical industry corresponds to the synthesis planner in the chemical industry. The corresponding tool is Computer Assisted Synthesis Planning (CASP). With it one discovers reaction paths and preliminary products that deserve consideration for the final product with the desired characteristics.

In order to bring new products quickly on the market designers from several firms (e.g., manufacturers of car bodies and of presses) at times work together simultaneously at a product without meeting in the same place (concurrent engineering). Here is a strong application potential for groupware (see section 4.3.1.4).

## 5.1.1.2 Computer Aided Planning (CAP)

*Computer Aided Planning* (CAP) implies the partially automatic development of work plans (manufacturing instructions) or—in opportune cases— entire manufacturing processes (Computer Aided Process Planning (CAPP)). The application system must derive the work plans from the geometry data and the bills of materials, as they come from the CAD (see section 5.1.1.1), and possibly from the already stored work plans of similar products.

*A PRACTICAL EXAMPLE*

*The company Carl Zeiss supports work plan generation for materials and individual parts by using an XPS. This work area was chosen since the problem was not solvable within a reasonable time frame with conventional software due to its complexity. The entire system consists of a strategy component that chooses the individual operations (e.g., separating raw parts from glass blocks, surface coating) and a classification component in which the target values are determined for each operation (e.g., tolerances). For this purpose extensive controlling know-how, which assures that the work processes are determined in the right sequence, was stored in the strategy component. The role of the explanation component is to provide the user with reasons for a chosen work process plan. Experiences with XPS have shown that in ideal situations up to 95% of the time needed to generate a work plan in the conventional way may be saved.*

## 5.1.2  Marketing and Sales

### 5.1.2.1    Customer Inquiry and Sales Offer Processing

Very powerful tools are available to handle sales offers. If they are combined one may speak of closed *sales offer systems* that support the sales force when visiting the customer. One also refers to this as *Computer Aided Selling* (CAS).

Imagine that the salesperson of a printing machine manufacturer wants to sell the owner of a private copy shop near a university various devices (copiers, sorting equipment). She carries a notebook with her. During the first phase she interviews the owner of the copy service shop by using a stored checklist. She records, e.g., the number of copies made each day, information about the quality requirements of customers, the average size of the lecture notes and term papers to be copied, etc.

With the use of an *electronic product catalogue* products meeting the needs of the copy shop are projected onto the screen. They appear as photographic pictures and also as diagram sketches. After a pre-selection by the customer the system puts together an arrangement of equipment during the *configuration process*. In doing so it examines many conditions and thus ensures that on the one hand the required technical performance (operational capacity, quality, maximum footprint) is delivered and on the other hand that the chosen components can be integrated.

While using a stored price list for the individual components the application system calculates in the next phase the offer price. Since this price is way too high for the customer, the system generates a leasing contract with low monthly payments. As a special service, the application system calculates the cost effectiveness of alternative machine configurations in the copy shop under differing assumptions. E.g., it may analyze that after the investment in the extra high capacity device twice as many new students may be acquired as customers or that sales may go down by 20% since the costs for using newly installed machines in the university library have been lowered. In the next step the offer is printed out and is stored on the notebook. Finally, the salesperson transmits the offer and potentially the order (e.g., using a wireless modem) to the central computer of her printing machine manufacturing company.

When selling products that have been individually manufactured for customers it may be necessary to involve several specialists and departments. This suggests that a Workflow Management System (WMS) (see section 4.3.1.2) may be useful.

*A PRACTICAL EXAMPLE*

*INA Wälzlager Schaeffler oHG deploys the WMS „BusinessFlow" in support of the business process "Customer Special Order Inquiry/Quotation Processing". A work flow component makes it possible to assemble all necessary documents (e.g., the product development request and customer drawings entered by scanner) in an electronic document folder which may then be transferred from employee to employee as needed. For the processing of the individual activities application systems are integrated: E.g., INCAS (Integrated Computer Aided Selling) or INDIOS (a program for the automatic generation of offers and quotations). The result of the process is either a customer-specific offer or the rejection of a customer inquiry [Mertens et al. 94].*

### 5.1.2.2    Quotation Follow-up

This program evaluates periodically the provided offers and quotations and makes available *offer and quotation reminders* to the sales department when needed. The appropriate salesperson will then contact the customer accordingly.

### 5.1.2.3    Order Entry and Order Verification

The order entry system is one of those points where many and important external data are entered into the information systems of an enterprise. Here we are concerned with efficient entry of information, but we also want to assure the accuracy of the information.

Aside from the conventional possibilities (entry of the orders received by mail, telephone, fax, etc. at a computer monitor) one has to strive above all for solutions with which data entry at the monitor does not apply and in which the information systems of the customer and the supplier are integrated as strongly as possible via EDI or XML. In that way the salesperson may fill out at the customer's premises a computer-readable receipt form or may enter the order into a mobile terminal. Possibly the customer uses the Internet to prepare and to transmit these order data.

*A PRACTICAL EXAMPLE*

*Ford-Werke AG in Germany has its central parts administration integrated with the ordering systems of the Ford dealers. The integrated system is called DARTS (Dealer Application Remote Terminal System). An order scheduling program running on a computer in a dealership makes suggestions when and in which amounts accessories and spare parts should be ordered. The dealer authorizes or modifies these suggestions and stores the final orders in the dealership's computer. Periodically the central DARTS computer fetches the orders automatically from the dealers' computers and transmits these collectively to the order processing programs in the central spare parts administration of Ford Motor Works in Cologne.*

Important components are the various *verification* features. They must assure that as few incorrect data as possible enter the integrated system. The *technical* verification examines whether the desired variant may be delivered or whether maybe an error occurred during the configuration by the salesperson or whether a mistake was made by the customer. For example, a medical device intended for export to Germany may be specified with an adaptor that works fine in the United States, but does not meet the specifications of the German electricity grid. During the *credit assessment* one checks whether after acceptance and delivery of the order the customer would exceed a credit limit such that payment may be jeopardized. The *due date verification module* checks whether the customer's desired delivery date may be kept. In order to do this we first will have to query the inventory status starting with the finished products, on to intermediate products and up to externally procured parts. If one does not have enough supplies then the system has to estimate whether the required production processes may finish on time.

The program complex order entry and order verification finishes its tasks with the printing of the order confirmation and the storing of the orders received as transfer data for the PPC system (see section 5.1.5.1).

## 5.1.2.4    Customer Relationship Management (CRM)

*Customer Relationship Management* is a customer-oriented approach that attempts to develop and tighten long-term, profitable customer relations via individual marketing, sales and service concepts through the use of modern information and communication technologies. The challenge for information systems in this area can be seen in the need to integrate (*horizontal integration*, see section 4.2) partial systems for the pre-sale, sale and post-sale phases (e.g., warranty and returns processing). But also *vertical integration* (see section 4.2) helps, since the captured information (e.g., in data warehouses, see section 3.2.2) serves to provide those responsible for marketing and sales with important information about preferences and behavior of customers. A few examples of concrete functions of CRM systems are:

1. Storage of characteristics of the customer's business and its contact persons such that knowledge is preserved even when salespeople quit

2. Updating of the customer relationship (what and when did the customer buy from us?)

3. Analysis of the customer data, e.g., through the use of database marketing or data mining (see section 4.3.2.2)

4. Suggestions for the sales department that certain campaigns or special offers are necessary, e.g., support when a trade exhibit is forthcoming or

when for equipment one year after installation a general overhaul is recommended

5.  Selection of holiday gifts that fit the customer's profile

*A PRACTICAL EXAMPLE*

*A manufacturer in the sporting goods business established three business areas: sport textiles, sport shoes and sport equipment. In the worst case, a large sporting goods store in Philadelphia is being visited on the same day by three salespersons from each of the three respective business areas. The information system examines the customer calling plan and points out the conflict.*

## 5.1.3  Procurement

### 5.1.3.1  Order Scheduling

Order scheduling, in principle, is concerned with the "programming" of the geometric presentation of figure 5.1.3.1/1.

First the system determines for each part the *safety stock* e. The entrepreneur determines the number of days $t_e$ during which he/she would like to remain ready for delivery, even when his own supplies are delayed due to disruptions (e.g., a strike at the supplier's). The application system multiplies $t_e$ with the observed average daily outward stock movement from the warehouse and thus calculates e. In more sophisticated versions e may be amplified through the use of statistical methods when the forecasting for outward stock movement is associated with large uncertainties (large differences between forecast and actual calculations, registered by the system itself).

For the prediction of outward stock movement one differentiates between program control and demand control. In the *program-controlled* calculation the demand for component assembly groups and parts is computed based on on the planned sales and the production program. This is addressed in section 5.1.5.2.

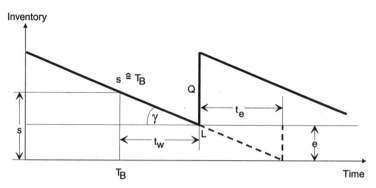

e    = minimum stock level
$t_e$   = safety time (to cover forecasting deviations and other
         uncertainties)
$t_w$   = replenishment lead time
$T_B$   = order due date
s    = quantitative order limit
Q    = lot size/order quantity
$\gamma$    = angle, reflecting the speed of stock withdrawal

*Fig. 5.1.3.1/1    Determination of Reorder Time and Amount*

With the *demand-controlled* forecast the information system observes the outward stock movement and thus determines future demand. Figure 5.1.3.1/2 shows as an example how input data concerning material movements may be created.

| Type of Material Movement | Affected Warehouse | Reported by Program (Section) |
|---|---|---|
| Delivery to customer | - Finished product inventory | Shipping logistics (5.1.6.3) or invoice processing (5.1.6.4) |
| Customer returns | + Finished product inventory | People-generated warehouse inflow notices (5.1.4.2) |
| Delivery by suppliers | + Raw material, outsourced parts | Incoming goods inspection test (5.1.3.4) |
| Goods returned to supplier | - Raw material, outsourced parts | People-generated warehouse outflow notices (5.1.3.4) |
| Inflow to and outflow from shops | +/- Shop inventory | Shop floor scheduling (5.1.5.7) |
| Inventory differences | +/- Miscellaneous warehouses | Inventory (5.1.4.3) |
| Warehouse transfer | +/- Miscellaneous warehouses | People-generated notices (5.1.4.2) |
| Legend: +: Inflow to, -: Outflow from | | |

*Fig. 5.1.3.1/2    Material Movement in Integrated Information Systems (A Simplified View)*

In many firms the core of the procedure is the first order exponential smoothing in accordance with the formula:

$$\overline{M}_i = \overline{M}_{i-1} + \alpha(M_{i-1} - \overline{M}_{i-1})$$

The symbols denote:

$\overline{M}_i$      = Predicted demand for period i

$\overline{M}_{i-1}$    = Predicted demand for period i-1

$M_{i-1}$     = Actual demand in period i-1

$\alpha$         = Smoothing parameter ($0 \leq \alpha \leq 1$)

The demand for period i is estimated by correcting the predicted value for the period i-1 by the fraction $\alpha$ of the thereby occurring prediction error. The size of $\alpha$ determines how sensitive the forecasting process will react on the newest observations. The smaller $\alpha$ is, the stronger the forecast values of the past are taken into account. Looking at the formula gives us an indication of this effect, e.g., if one assigns $\alpha$ to be as small as possible, i.e. zero: Now the system picks the old forecast value also as the new predicted value, i.e. the last observation $M_{i-1}$ does no longer play any role. For more complicated demand processes (cyclical trend, seasonal dependencies, overlap of trend and season or demand thrusts through sales campaigns) the exponential smoothing has to be enhanced.

The order scheduling program finds the intersection L (desired delivery date) of the outward stock movement line with the parallel to the x axis that marks the safety stock e, and moves to the left from this point by the amount of the reorder time $t_w$. In doing so the order date $T_B$ is specified. This is the point in time when an order will have to be placed such that after the reorder time period has passed the ordered parts will arrive on time. The x-coordinate value $T_B$ corresponds to the y-coordinate s. This is the reorder level.

In the next step a low-cost order quantity Q is calculated. If the part is obtained through in-house production, then the order for a finished product will have to be transferred to the application system *Primary Requirements Planning* (see section 5.1.5.2). For an intermediate good or single part the transfer data will need to be sent to the application system *Material Requirements Planning* (see section 5.1.5.3). If we are dealing with an external procurement, however, then material scheduling will pass along the data to the purchasing system.

### 5.1.3.2    Purchasing

One needs to distinguish whether a) only one supplier is under consideration, b) the systems may determine a supplier on its own (in a module supplier selection) or c) the choice becomes the duty of the purchasing officer. In the cases a) and b) the computer can print the order and store it in a temporary reservation file. In case c) the purchasing officer is presented with the choice of suppliers on the monitor. The orders may only be triggered when a human being has entered his/her decision into the computer. In some firms the application system transmits its orders, e.g., via the Extranet to the computer of the supplier (interorganizational integration or supply chain management, respectively, see section 5.4).

During the classical procurement process the ordering entity, e.g., a raw material warehouse, enters an order request into the system. This request is processed at the PC of an employee within the purchasing department (checking against the purchasing budget, choice of supplier, where appropriate a modification may be made in the computer-suggested lot size) and is then passed along to the supplier. In case of the so-called desktop purchasing in our example the warehouse scheduler places the order without the participation of the purchasing department. The rules via which the purchasing department employee checked the order request are now depicted in an electronic program. As long as there is no interference the computer of the warehouse scheduler transmits the order in electronic form to the supplier.

The purchasing or procurement function is according to many observers one of the partial systems of a manufacturing firm that are drastically changed by the Internet ("electronic procurement"). We may distinguish among various levels [Mertens 01, pp. 94-95]:

1.  *Information collection from WWW presentations:* The purchaser acquires on the Internet the newest product descriptions and prices, investigates potential suppliers, etc.

2.  *Internet-Shops:* Buyers visit virtual show rooms on the Internet and order directly.

3.  *Internet-Shopping-Malls:* Here several Internet shops are brought together in a portal. The buyer may thus inquire about the availability and conditions of a multitude of potential suppliers.

4.  *Internet Request for Proposals:* The buyer may post his needs on an Internet portal. An intermediary compares the specification with the tender of potential suppliers und transfers the inquiry with the request to make an offer.

5.  *Internet Market Places:* One may see them as a further development of Internet requests for proposals. The main difference is that the potential suppliers after the request for proposals have the possibility to examine the offers of the other competitors and to improve their own in favor of the buyer without interference by the market place operator (online purchase auction).

### 5.1.3.3   Delivery Monitoring

The program "order monitoring" controls in regular intervals the "external procurement orders", created by the program "order disposition". If delivery due dates are surpassed, then reminders are sent to the suppliers. Each occurring reminder is registered in the supplier master data. For example, the increase in the total number of reminders may influence the supplier choice during the next purchasing cycle.

### 5.1.3.4  Goods Receiving Control

Subjects of the goods receiving control are the *quantity* and *quality control*. The application system is supplied with the goods receiving information that, e.g., the order schedule program produces. The goods receiving control may thus recognize whether or not a delivery was ordered at all and whether the delivered quantities are in agreement with ordered quantities.

An elegant possibility to streamline and to make *quality control* more sophisticated through the use of information systems is the use of the "dynamic" sampling procedures [Mertens 01, pp. 110-112].

*A PRACTICAL EXAMPLE*

*In several European IBM plants the quality management system SAP R/3 QM is used. This system operates under varying rules and data. For example, the dynamic sampling rule contains inspection stages (intensified – normal – reduced – skip (see below)) and the conditions for a change among them. These changes occur depending on the inspection results of the individual inspection lots and on the acceptance/rejection decision. The rules determine whether or not the inspection volume is to be reduced or increased (tightening or widening the sampling net) or, respectively, at what level of error the lot is rejected. For example, the system switches from reduced to normal control as soon as one faulty part out of four units in a lot sample is reported. Depending on the stored rules and data SAP R/3 QM makes the sampling volume dynamic between 100% and 0% (inspection waiver, skip). Skip implies that a certain number of future lots or characteristics are not being examined.*

## 5.1.4  Warehousing

### 5.1.4.1  Material Valuation

The application system extracts valuation approaches from the material master data such as (with externally procured material) the prices from the order or delivery, permanently stored stock prices or the newest costs based on a product costing analysis (see section 5.1.9.1). One may also consider a simple evaluation calculation such as with smoothened averages (with each inward flow an average price is calculated).

### 5.1.4.2  Inventory Control

Mechanized inventory control is in principle very easy. One follows the following formula:

New inventory = old inventory + inflows - outflows

Complications, however, may arise, e.g.:

- Aside from the 'bureaucratically' managed, i.e. using stock requisitions and delivery notes, warehouses there is shop floor stock for which not every single movement is accompanied by a book entry.

- One has to consider reservations, i.e. parts, that are physically still in the warehouse even though they are already encumbered and may only be delivered for a specific purpose.

### 5.1.4.3    Inventory

Stock-taking as an example demonstrates how information systems trigger processes carried out by men and thus makes essential contributions to the orderliness of companies (concurrence between "book inventory balance" and actual "inventory"). Occasions requiring inventories that applications systems are capable of determining themselves are [Mertens 01, pp. 122-124]:

1. Exceeding of set maximum stock levels

2. Falling below a minimum level (it is recommendable to conduct an inventory when few items are in stock because then the effort to count is small)

3. The formation of book inventories below zero

4. Within a parts or item category a certain number of fluctuations has occurred (with this a certain probability exists that an error occurred during the postings)

5. With a parts or item category no fluctuations were observable over a longer period of time (possibly this parts or item category does no longer exist)

6. Control via computer-generated random numbers to ensure the so-called surprise effect when there is risk of theft

7. Activation on an appointed date

The inventory is conducted either by counting everything or through sampling. During the sampling inventory the information system determines a suitable *sample size* using methods of mathematical statistics and extrapolates the counted results.

### 5.1.4.4    Support of Processes in the Warehouse

When combining business administration processes with technical information systems many opportunities for efficient warehouse management arise. Among them are:

1.  Information systems administer high rack storage areas. Palettes are being transported automatically to available storage locations using horizontal and vertical movement. The individual palettes are not sorted based on any particular order (so-called *random* or *chaotic storage*). Since the computer stores an image of the warehouse, the system is capable of finding available positions any time.

2.  During stock withdrawal "positions" as part of a *consignment* (order or shipping process) are fetched (partially) automatically from their storage location, sorted and are being transported to the location for packing and shipping.

*A PRACTICAL EXAMPLE*

*Avon Cosmetics™ opens the shipping cartons that had arrived folded with an automated machine. Also automatically each carton is marked with a bar code that identifies it as part of a customer order. A control system reads this bar code and delivers the carton over belts, gates and switches only to those unloading/loading locations where there are ordered parts/items. The employees handling the consignment receive a graphical presentation on a monitor that shows to which shelf location and space they have to move. This shelf space is being illuminated at the right moment and on a digital display the employee sees how many units (e.g., lip sticks, tubes) will have to be picked and be placed into the carton. After this is done the employee pushes a button to report the process as completed. The carton is then transported automatically to the next location. In that way a sort of "consignment progress control" occurs. The system transmits data to a logistics manager who is then informed of which shipping tasks are forthcoming and how he/she is to manage these shipments [Haberl 96].*

## 5.1.5  Production

### 5.1.5.1  CIM–The Complexity Problem

Information processing in the production area is earmarked in that business management data processing, technical data processing, as well as physical production processes have to be integrated. The underlying concept is labeled *Computer Integrated Manufacturing* (CIM).

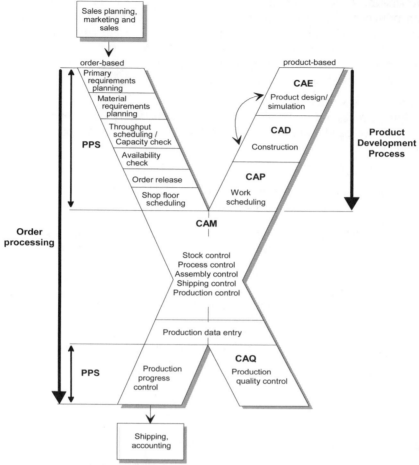

*Fig. 5.1.5.1/1    The CIM Concept*

On the business management side (in the narrow sense) we have to manage production planning and control (PPC). Plant maintenance planning and control are omitted here. PPC constitutes the transaction chain in the flow of orders. The transaction chain that connects the design, the physical manufacturing and the quality control of the product is linked by the so-called C techniques. At the cross-over point *Computer Aided Manufacturing* (CAM) the two chains are so intertwined that a separation into the business management and technical parts is barely recognizable. Figure 5.1.5.1/1 is to illustrate this situation [Scheer 90, p. 2].

This illustration is of relevance mainly to firms, e.g., in the mechanical engineering industry, that manufacture products that have been ordered by customers on an individual basis or with certain variants in job-shop manufacturing settings. For firms that turn out mainly mass-produced items for an "anonymous" market, such as cleaning detergent, other configurations of the building blocks will have to be chosen.

A very difficult problem in the conceptualization of application systems in the production area is the intensive reciprocal interaction among the individual systems. Basically, one should depict CIM as one huge *simultaneous model*. Figure 5.1.5.1/2 shows just one example of the effects of integration: If one increases the lot size, this implies that one is willing to put up with increased inventories, that tie up more capital. However due to shorter set-up times, bottlenecks are better utilized. Thereby the throughput times of orders decline initially during the manufacturing process. After exceeding a minimal value, cycle times will tend to increase again, because lots will increasingly have to wait directly at the manufacturing unit where a larger and previously scheduled lot is being processed for a longer period of time.

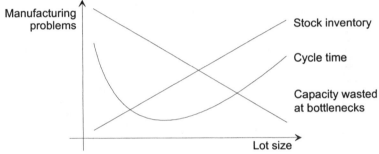

*Fig. 5.1.5.1/2    Integration Effect in Manufacturing*

The production flow depends on various factors, e.g., lotsize at various manufacturing levels, production sequences, as well as the choice of constructive variants and alternatives in conjunction with work flow, plans and schedules. Even with the largest available computers today it is still impossible to master a simultaneous optimization. During the course of several decades a sequence of the modules established itself that was considered practical in many respects, but not always optimal. It is based on the following considerations.

### 5.1.5.2    Primary Requirements Planning/MRP II

Primary demand planning balances roughly desired sales or production quantities, respectively, with available manufacturing capacities. This early-on coordination of available capacity and capacity requirements is to avoid that the floor shop is inundated with unrealistically planned production orders. For this one may use, e.g., sales forecasts based on statistics [Mertens 94], as well as the matrix model outlined in section 5.1.12.

These systems are referred to as *Manufacturing Resource Planning* (MRP II). We need to distinguish between MRP II and MRP I "Material Requirements Planning" (see following section). In addition, MRP II concepts are characterized by a multitude of feedback loops, but these are not addressed here.

### 5.1.5.3    Material Requirements Planning/MRP I

The demand for final products (derived from order entry, sales planning or primary demand planning) must be broken down ("bill of materials explosion") into its components (*secondary demand*). Figure 5.1.5.3/1 shows such a bill of materials.

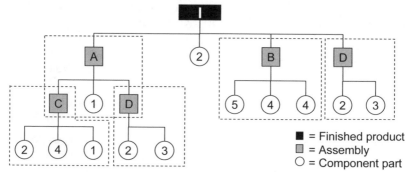

■ = Finished product
▢ = Assembly
○ = Component part

*Fig. 5.1.5.3/1    Manufacturing Structure of a Product*

It is organized as a building block parts list, i.e. one recognizes (in the shape of the dashed rectangles) from which subordinate parts each respective supra-ordinate part is assembled. The application system, e.g., determines that per passenger car five wheels are needed, including the spare. The assembly team "wheel" would subdivide this again into one rim, one tire and four clamping bolts, etc. In that way one would calculate first the gross requirements of the assembly teams and the component parts. Comparing these to the available inventory yields the net requirements. The application system examines also whether or not cost-effective bulk orders may be placed through bundling of demands for different future time periods. In several industries, e.g., aluminum, glass, optics, paper and textile, we may specify through rather complicated mathematical procedures how parts (e.g., paper webs) will have to be cut out of larger raw material units (e.g., paper master rolls) such that waste is minimized. During such materials demand planning the transition to the subsequent time scheduling is implemented in that the system considers the associated lead time. This is the time span by which the subordinate component has to be available earlier than the supra-ordinate component such that the subsequently requisitioned parts may be assembled in a timely fashion. The results of the procedure are roughly planned work, manufacturing or production orders, respectively, or (with external procurement) requirements to be transferred to order scheduling (see section 5.1.3.1).

## 5.1.5.4    Throughput Scheduling

Whereas volume planning determined availability scheduling, i.e. those points in time at which a part is to be delivered, through forward-shifting, throughput scheduling has to generate the *starting dates* of the individual work processes.

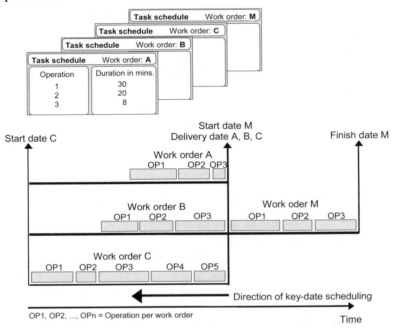

*Fig. 5.1.5.4/1    Backward Scheduling*

A method for this is backward scheduling that calculates from the delivery deadline found in material requirements planning toward the present. An example in figure 5.1.5.4/1 is the work order M in assembly requiring the finished products of the work orders A, B and C. One should note that in this phase waiting periods, as they tend to come about through capacity bottlenecks, are not considered. In other words: We work here with the simplified assumption "capacity is infinite". With other, more complicated and computation-intensive methods this simplification is not applied, as they conduct simultaneous scheduling and capacity planning.

Particularities occur when the application system determines that a work process should have started several days or weeks ago "prior to the present" (sometimes students gain such knowledge as well during the manufacture of the product "examination"!). In order to avoid revising the hitherto production planning because of the later delivery date, the application system will attempt to shorten the throughput time in comparison to the planned values. For example, it may test whether several machines are available for a work process and then split a lot onto two or more manufacturing resources that then "divide" the work among themselves. The system then has to determine through the use of parameters defined by manufacturing management the trade-off between lead-time reduction and the additional expenses for setting up a second, third, etc. machine.

### 5.1.5.5    Capacity Balancing

If in the throughput scheduling no attention had been paid to capacities (see above), it may occur that in individual periods certain work places are heavily overloaded whereas others are underutilized (cf., fig. 5.1.5.5/1).

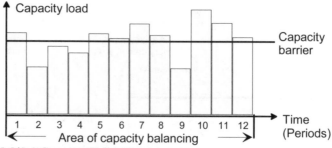

*Fig. 5.1.5.5/1    Capacity Balancing*

Here capacity balancing is put into place. You recognize at first glance that, e.g., one needs to tip the peak in period 10 into the valley of period 9. People recognize this due to their pattern recognition capabilities in which they are clearly superior to a computer. This is why in many cases one will not attempt to automate capacity balancing, but to show the capacity "mountains" on the monitor of a *control station* called *Leitstand* (see section 5.1.5.8) and to deliver appropriate information about which individual production and customer orders contribute to the work-load during a particular time period. Based on this information a manager can decide on the rescheduling measures.

### 5.1.5.6    Availability Check

It would be awkward if a computer triggered the start of a work order that could not be executed because during the same time period a needed machine is down for maintenance or repair, externally procured material did not arrive in a timely fashion due to tardiness of the supplier or a control program (NC-program) has not been written yet. Maybe employees with the right qualifications are on holiday on that particular day. It is the goal of an availability check to set aside such production orders for which any resources might be missing (compare the regarding Available To Promise (ATP) technology in section 5.4).

### 5.1.5.7 Order Clearance

The order clearance process chooses (while being controlled by parameters) a subset of those orders which have passed the availability check for actual production. For example, all work orders are selected that have to start within the time span "clearance day + ten working days" according to throughput scheduling.

### 5.1.5.8 Job Shop Scheduling

One of the tasks of floor shop scheduling is to find a machining sequence for the orders at a work place that fulfills certain goals as well as possible. Such goals may be the *minimal overall lead time* of the lots, *maximal capacity balancing, minimal capital commitment, minimal set-up costs, maximal adherence to schedules* or also *simplicity of control processes*. Since the emphasis on goals may vary greatly in various industries, in different strategic circumstances or varying economic conditions very complex control tasks arise.

Control approaches may be structured according to whether the next lot to be machined at equipment that just was freed up is to be determined or whether it is more important to assign appropriate manufacturing resources to upcoming production orders when there are several to choose from. The latter plays a role in rolling mills or also in the paper industry and here especially with regard to cutting problems (see section 4.3.2.4).

By using *priority rules* it becomes possible to consider the actual emphasis of the different goals. One may apply, e.g., a computer-based control system such that at a bottleneck the system chooses the very lot among a set of waiting ones that fits best for the actual set-up condition of the manufacturing resources. In this way it is possible to keep the set-up costs low. On the other hand, if one decides to prioritize the lot that ties up the largest amount of capital, it is possible to guide capital-intensive products through the shop in shorter time and as a result reduce the tying-up of capital.

When a largely automatic control is not (yet) possible due to the outlined complexity, often *production control stations* (so-called Leitstands) are utilized that support scheduling by a person. With suitable user interfaces, control station personnel will be able to evaluate the current manufacturing situation (e.g., capacity balancing of bottlenecks, idle machines, late orders, minimum inventory levels, resources not available (see section 5.1.5.6)).

Shop floor control issues documents necessary for production (job tickets, time tickets, material documents, quality control receipts, etc.). It is practical to make these machine-readable (e.g., in that they feature a magnetic strip), since when they return from the manufacturing site they may be read into the computer system again ("return data carrier") (see section 5.1.5.11). Some firms do not issue such documents; instead employees identify on their screens what they should produce and how ("paperless factory").

### 5.1.5.9    Computer Aided Manufacturing

The term Computer Aided Manufacturing (CAM) does not only comprise the information systems support of physical production in the narrow sense, but also systems serving the automation of the functions *transporting, warehousing, testing* and *packing.* CAM administers *numerically controlled machines* (CNC, DNC machines), as well as *manufacturing cells* and *flexible manufacturing systems, processes* (e.g., in the chemical industry), *robots* (RC = Robot Control) and various types of *transportation systems.* One might add the *administration of warehouses,* especially buffer warehouses in manufacturing.

The position of CAM as a manufacturing information system in the manufacturing execution is characterized by two attributes [Mertens 01, pp. 182-187]:

1. One attempts to accompany the material flow with CAM over several phases (cf., fig. 5.1.5.9/1). A comprehensive CAM system sets up manufacturing resources with tools automatically, captures their down times and machining times, recognizes worn-out or defective tools and exchanges these. Moreover, work pieces or the material are taken from the warehouses in accordance with the production schedules. Next they are passed on to the manufacturing resources in a suitable sequence (e.g., in a flexible manufacturing system such that as few set-up processes as possible are required) and the physical manufacturing processes are controlled (e.g., the setting of a spot weld by a robot or the drilling speed of a drill).

   Beyond that *driverless transportation systems* are directed, finished products are packaged and made available for shipment. In that way one arrives at the *"sparsely populated factory"* in which people carry out merely controlling tasks. Coordination is often the duty of a host computer or a *production control system* consisting of several networked computers. A control system as a CAM component may not be seen as comparable to a control station (*Leitstand*) as part of a production planning and control system (see section 5.1.5.8).

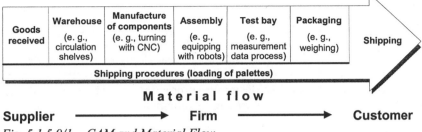

Fig. 5.1.5.9/1   *CAM and Material Flow*

2. Not only with the embedding of CAM into the information architecture of the manufacturing firm (cf., fig. 5.1.5.1/1), but also internal within the CAM complex one often constructs multi-level hierarchies. The example in figure 5.1.5.9/2 depicts the simplified information systems architecture of the BMW™ Group.

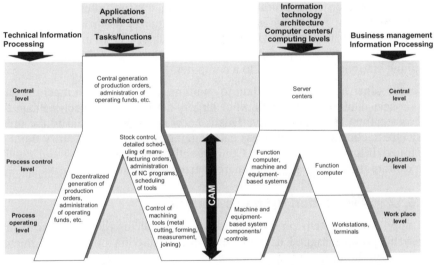

Fig. 5.1.5.9/2   *Hierarchies within the CAM Complex/Information Architecture*

Essential for the implementation of CAM is an intensive information transfer among computers on various hierarchical levels and manufacturing, transportation, warehousing, as well as possibly testing activity clusters.

## 5.1.5.10   Computer Aided Quality Assurance (CAQ)

The assurance of production quality is often also described with the term *Computer Aided Quality Assurance* (CAQ). An extended meaning of CAQ includes also the control of product quality during the design phase (see section 5.1.1.1), quality inspection in the goods receiving department (see section 5.1.3.4), maintenance or repair of the shipped equipment or machines at the customer's and the processing of complaints. Here, one gets closer to the term Total Quality Management (TQM).

Modern solutions make it possible that an application system triggers *tests* (e.g., electrical measurements, surface tests, physical, chemical or microbiological inspections). If not all products are tested continuously and the computer tests are based on *sampling*, one achieves an economization or streamlining and, also the possibly desirable surprise effect. This keeps working staff alert, as it is not possible for them to anticipate any control activities.

### 5.1.5.11    Production Data Entry

In the *production data entry* messages returning from production processes (e.g., time, quantity, pay, material withdrawal, and quality control data) are entered into the computer and are stored with preliminary reservation data of the placed production orders. One of the challenges in the further development of production data entry systems is to capture as much data as possible automatically, e.g., from production equipment, transportation equipment or automated testing machines (*machine data logging*) or directly emanating from a process (*process data logging*). It is possible, e.g., in a pharmaceutical company to register the quantity of the manufactured granulate material at a weighing station that is coupled to a computer.

On the other hand it is important to examine, especially with machine data and process data logging, the flood of largely automatically received data for their correctness and plausibility. This is necessary since, similar to the order entry (see section 5.1.2.3), process data logging is an important entry point to integrated information systems. Consequently, errors occurring during the data capture may easily trigger numerous erroneous after-effects.

### 5.1.5.12    Production Progress Control

The application system production monitoring utilizes the production, machine and process logged data to recognize production progress. If missed deadlines are looming, it issues *reminders*.

## 5.1.6  Shipping

### 5.1.6.1    Assignation

Assignment decisions are necessary when no complete customer order manufacturing exists from the lowest initial production stage. In the simplest case the application system sorts in each run the newly arrived customer orders based on delivery dates and within the delivery dates based on priorities depending on customer- or order-size. Then the upcoming customer orders are fulfilled as long as the available supply lasts [Mertens 01, pp. 214-217].

For many practical purposes this simple procedure in not sufficient. For example, for especially important customers stock inventory has to be reserved even when the delivery is due much later. Moreover, one has to avoid that a very small number of units (e.g., with a knit ware manufacturer: for only individual sizes of a sweater in particular colors) is assigned when the complementing order items will not be completed in the production department in the near future. For such assignment programs the information systems specialist will have to develop possibly a complicated body of rules and regulations.

### 5.1.6.2    Delivery Clearance

It is the duty of the application system to examine whether the items to be delivered as determined by the assignment program are ready for shipping. This is simple when the customer order is completely ready prior to the confirmed delivery date. It is more difficult for such a scheduling system when only a portion of the customer order is ready for shipment at the latest possible delivery date and a decision must be made whether partial deliveries may and should be expedited. Such programmable rules, for example, are [Mertens 01, pp. 218-220]:

1.  Partial shipment occurs as soon as a percentage p of the order is ready for shipping (it would not make sense, e.g., when a company that manufactures photo-technical papers would ship five pieces that were discovered by chance in an already opened box as a partial delivery of the customer order for 2,000 pieces).

2.  A shipment that strictly speaking is only a partial delivery is treated as a complete fulfillment of the customer's order when the remaining amount is $\leq r\%$.

3.  Whether or not a partial order shipment occurs is determined by using a model that considers the data constellation within the shipping logistics area, such as the balancing of shipping capacities.

Cleared deliveries are transferred as preliminarily reservation data to the shipping logistics area.

### 5.1.6.3    Shipping Logistics

Shipping logistics encounters many complicated scheduling problems that manifest themselves quite differently in various industries. A complicated case exists when at the same time customers are to be supplied and off-premise warehouses need to be replenished with reserve stock. Complicating matters further, the source for delivery may be several production sites and/or central warehouses and, finally, one has to choose among several transportation modes (e.g., truck, airplane, ship) and logistics service providers (e.g., the firm's own truck fleet, shipping company, postal service or parcel delivery firms). Even though this is a problem complex suggesting the application of simultaneous optimization routines, in reality one has to employ successively individual program modules and one has to live with sub-optimal solutions. Such modules are:

1. Selection among available warehouses for the shipment
2. Bundling of shipments to a given load
3. Determination of trip route
4. Assigning of the route to vehicles and drivers

The further development of partially automated production shipping logistics is supply chain management (see section 5.4).

### 5.1.6.4    Invoicing

The invoicing program generates the customer invoices based on the order and shipping data and using both the customer and materials master data. In this process the various deductions (e.g., discounts, incentives) and mark-ups (e.g., for small orders, packaging, shipping) need to be considered.

## 5.1.7  Customer Service

### 5.1.7.1    Maintenance/Repair

Among others, a system for the support of maintenance and repair includes the following functions:

1. Monitoring of due dates for preventive maintenance
2. Computer-supported assistance for diagnosing the condition of shipped products (in suitable settings it may be possible to conduct remote diagnostics while using data transmissions)
3. Selection of suitable field staff based on special knowledge, actual workload and geographic proximity; possibly they will be informed by the computer over the Internet

Related to this topic are help desk systems: The "helper" in the manufacturing firm discusses the fault report with the customer by phone, asks for the observed symptoms and enters these into the computer. At the same time the computer displays the particular characteristics of that machine (electronic product description) after retrieving this information from its memory and recommends ways to eliminate the error with the help of a knowledge-based system.

### 5.1.7.2    Customer Queries

With customer queries the danger exists that the complaint moved from office to office and that the customer thus will have to wait a long time until a response is received. If the customer complains, e.g., that a defective machine was delivered, it may be that responsible employees from purchasing, manufacturing, shipping, legal department and accounting (the latter department because financial reserves may have to be created) have to be asked for their assessment. In such cases it may be advisable to deploy workflow management systems (see section 4.3.1.2). Moreover, knowledge management systems in which technical solutions with similar problems of the past are described may offer valuable assistance.

## 5.1.8 Finance

In comparison to other areas we find relatively few administration and scheduling systems in the finance area (excluding accounting).

An important, although difficult task, is the finance and liquidity scheduling. Here we are concerned with predicting the likely revenues and outlays and, depending on the accounting balance, deciding about use of available funds or about getting a short term loan. Especially international firms utilize a *cash management system* (see section 5.2.8.5) for this purpose.

Above all it is the duty of the computer to predict the extents and dates of payments. In integrated information systems we may draw, e.g., upon the following data: sales plan, goods on order, accounts receivable, accounts payable, purchase commitments, costs projections, regularly recurring payments (e.g., wage and salary payments or rents) and investment plan.

## 5.1.9 Accounting

### 5.1.9.1 Cost and Result Accounting

#### *5.1.9.1.1 Cost Center Accounting*

Information systems-supported cost accounting is largely a recreation of the procedure as it is conducted manually by the employees. For the machine hourly rate we have the advantage that in the CIM conception machine times may be determined easily and rationally through the integration with programs in the production area (see section 5.1.5.1). Similar things may be said about target costs that, e.g., may be calculated by multiplying the actual times (registered exactly in the production data entry (see section 5.1.5.11)) with the target prices. Finally, integrated information systems make it possible to determine aside from cost deviations also deviations of consumed quantities, as well as capacities and thus being able to deliver suggestions for an interpretation for the target vs. actual deviation. Altogether integrated information systems enable cost center accounting to work with very differentiated data. Those, in turn, make it possible to apply individual direct cost and product profitability calculations.

## 5.1.9.1.2    Product Cost Accounting

■  Preliminary Costing

For *preliminary costing* three data groups are at our disposal within an integrated concept:

1. Material master data
2. Parts lists or bills of materials
3. Task schedules with the work processes, manufacturing resources to be utilized and the corresponding times (component routing)

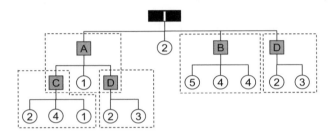

**Calculation procedure:**

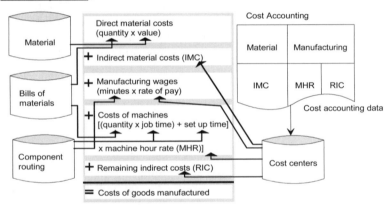

*Fig. 5.1.9.1.2/1 Preliminary Costing*

The preliminary costing program passes "bottom-up" through the parts list, from the component part to the finished product, and merges component by component. This is done while utilizing the latest actual costs per activity output unit (e.g., per manufacturing minute) which have been determined by the cost center accounting program. Figure 5.1.9.1.2/1 depicts this procedure schematically.

■  Post Costing Analysis

The *post costing analysis program* also uses many data from the production area. Data pertaining to material movement and time tickets or wage slips (delivered from the production data entry and pay accounting in machine-readable form) generate the direct costs that are posted to the cost object accounts. As far as one wants to consider indirect costs, if at all, this occurs through adding overhead charges to the direct costs and by costing the usage times registered by the production data entry system (e.g., via machine hourly rates).

### 5.1.9.2    Supplier Accounts Auditing

Supplier accounts auditing is another good example for how most of the data are made available in machine-readable format in an integrated information system, namely [Mertens 01, pp. 262-263]: The prices in the material master records, supplier-dependent conditions in the supplier master records, the *order* and *goods receiving* data in the preliminary data registers and the entered supplier invoices or the electronically transmitted ANSI X12 or EDIFACT invoices.

The application system is capable of conducting a number of checks with the preliminary registered data, especially on:

1. Concurrence between delivered and invoiced quantities
2. Concurrence between prices and conditions of the offers and their delivery
3. Correctness of the supplier invoice

In case of differences the system dispatches a message to the responsible employee who then will address the discrepancy. If problems cannot be solved, he/she would block the invoice and trigger further clearing.

### 5.1.9.3    General Accounting

The structure of general accounting programs is determined by the method of double book keeping. A large amount of the input data is being delivered by other programs, including, e.g., condensed accounting records from accounts receivable and accounts payable programs or material postings from the material valuation program.

Characteristic for an integrated information system are the excellent *reconciliation capabilities* that enhance the security in accounting (e.g., between general accounting and sub-ledger accounting or between receivable accounts of billing and the sum of the *debitors'* postings).

But even the input of booking procedures that still need to be handled manually has been streamlined, for example in that the accountant is being guided from posting to posting and thus is being made aware of any erroneous entries right away.

## 5.1.9.4    Sub-Ledger Accounting

### *5.1.9.4.1    Accounts Receivable*

Accounts receivable manages the preliminary data registry debitors. Business transactions are transmitted as transfer data by the billing program and are then posted by the application system into the receivable accounts.

When due dates are exceeded ready-for-shipment *reminders* are issued through the use of stored text modules. In accordance with the respective reminder level (e.g., first and second reminder) the program applies formulations of varying degrees of "sternness". Received *customer payments* are registered and the corresponding preliminary data registries are erased.

### *5.1.9.4.2    Accounts Payable*

The accounts payable program is rather similar to the accounts receivable program. A module, however, is to be considered through which *payments may be processed at an optimal point in time.* Here it is appropriate to build in a parameter through which management may provide general guidelines pertaining to the conduct of payments. Then it is possible to pay with or without cash discounts depending on a firm's liquidity.

## 5.1.10 Human Resources

### 5.1.10.1    Work Schedule Management

By using information systems it becomes possible to capture *time sheet* data efficiently and very accurately. The two main demands for *flextime systems* "sufficient information of the employee about the status of his/her time sheet account" and "transfer of the time sheet data into payroll accounting" are thus fulfilled much easier. Such modules gain increasing importance as firms introduce more and more multifaceted work schedule models.

*A PRACTICAL EXAMPLE*

*The employee carries a machine-readable company identification card that features a magnetic strip or even a chip. This will be entered into a card reader when he/she arrives for work and when he/she leaves. The electronic system reads the individual's identification number and stores it with the correct time. At the same time the system checks the arrives-leaves cycle and makes the employee aware of any discrepancies. The latter may be the case, e.g., when the employee forgot to enter his "leaves" data to the information system the prior evening. At the same time it is possible to display to the employee the accumulated total work time during a given month, as well as the target/actual comparison.*

## 5.1.10.2    Payroll Accounting

Under the label payroll accounting we may group the programs for payroll (for wages and salaries), training and development allowance, and commission accounting. Their structure may be determined by wage agreements and legal stipulations.

Among the tasks of payroll accounting programs is the determination of the gross pay based on activity and time sheet data or quantitative performance, marginal return or sales performance (as with commission accounting), the determination of *extra pay*, such as on holidays, the computing of *net pay* and *net salaries* while considering *taxes*, *social insurance contributions* and other deductions. Moreover, such programs also take into account the determination of *payroll deductions* such as the gradual paying back of pay advances or personal loans from the employer. Difficulties with payroll accounting programs are not so much with their development but rather with the ongoing changes that are often determined late by the legislator, especially concerning the tax laws. Consequently, the support and care for such systems is rather expensive.

## 5.1.10.3    Reporting Programs

Within the human resources area a multitude of notifications, in part due to legal reporting requirements, have to be made by the employer. Often such reports are just printouts of certain fields within personnel data bases. Examples are various reports pertaining to employment statistics and notifications about wage and salary changes to all employees.

## 5.1.10.4    Special Action Programs

Shortly before the due date for special actions or measures (e.g., medical routine examinations, a company-wide flu-shot clinic, expiration of work permits of foreign nationals) the computer issues notifications based on information stored in human resource records. The computer may place these notifications into the respective employees' mail boxes.

## 5.1.10.5    Employees-Tasks-Assignments

Such scheduling systems devise plans that determine which work place is to be occupied at what times by which employee. It is characteristic that, aside from the sometimes difficult to measure job requirements and the equally difficult to quantify human qualifications, many conditions will have to be met. Among these are labor laws and union bargaining agreements (e.g., arrangements for and duration of breaks).

The most powerful application systems use XPS techniques (see section 4.3.2.3) and query master data for employees and production equipment as well as stored working time models.

## 5.1.11 Facility Management

The term *facility management* means the computer-supported control and monitoring of building complexes. Sometimes one refers also to *intelligent buildings*. Mainly, we are here concerned with the *control of physical processes*. For example, the electronic system calculates suitable air temperatures for air conditioners to avoid overheating (energy management).

There are as well connecting points to business information systems. *Access control systems* for buildings or sections of buildings requiring some form of security should be mentioned in this context. If a fire alarm were to be triggered, a building management system prints out situation-specific roadmaps such that firefighting could be facilitated. Databases describing the layout and occupancy of offices (together with their equipment and furnishings, such as PC, communication technology or facilities for disabled persons) are helpful during moving or other reorganization efforts.

## 5.1.12 An Example of a Computer-supported Planning System

As an example of a computer-supported planning system we have chosen integrated sales and production planning. Following we will sketch particularly how the system is embedded into integrated information systems. We assume that a preliminary sales plan has already been established (possibly with computer support [Mertens/Griese 02, pp. 219-223]). Next we need to examine the planned sales volume and juxtapose it to the production capacities (see also the discussion on primary demand planning in section 5.1.5.2).

To do so, the computer-supported planning system takes the production guidelines from the work schedule data for all in-house produced parts and transforms this information into the capacity demand matrix (preliminary stage) (cf., fig. 5.1.12/1). In this figure columns represent the individual parts and rows the manufacturing resources, as well as the work places. The matrix elements contain the required time needed for the production of a unit of the part with the corresponding manufacturing resource (or the work place, respectively). By using parts lists a similar bridge or aggregation program can generate the composite demand matrix from the structure of the finished products. Since assembly processes require capacity as well, they will need to be defined as fictitious parts. By multiplying the two matrices an additional capacity demand matrix is being generated with columns listing finished products and rows depicting manufacturing resources and work places. The elements are now the capacity loads of the manufacturing resources and work places required for the production of one unit of the finished product. One should note that by multiplying the matrices the individual parts were "cancelled out". This process is schematically depicted in figure 5.1.12/1. Through the multiplication of the last matrix with the vector of the sales program one migrates from the one-unit-of-finished product perspective to the entire sales plan and thus receives the capacities that are required for the realization of the previous sales plan when considering individual manufacturing resources or manual work places.

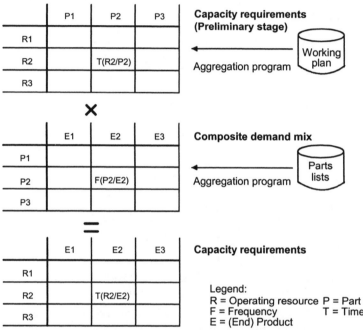

*Fig. 5.1.12/1    Determination of the Capacity Requirements Matrix*

If limits are exceeded substantially or shortfalls should occur, then alternative calculations or plans (see section 4.3.2.1) will have to be evaluated. In doing so sales figures, as well as production capacities will need to be changed (production capacities may be adapted to the sales plan, e.g., by investments in or disinvestments of individual manufacturing resources, by additional work shifts and by overtime or also outsourcing an order).

After a number of steps, including intensive human-computer interaction we assume that a sales plan was developed that can be realized based on the manufacturing capacities. Information systems then will have to suggest how the sales figures developed at the enterprise level for the individual products will have to be broken down appropriately for the lower levels (e.g., sales region/territory and then districts). In this the computer, e.g., may utilize the same proportions observed in the previous period. The results are sales plans for individual products in the districts and thus constitute planned targets for the sales representatives. These are then stored in the master data of the sales representatives and become the basis for actual/target comparisons, for the calculation of sales or for an incentive plan.

## 5.1.13 Example of a Computer-supported Control System

As an example of a control system we chose sales monitoring. We may see the present discussion as an extension of the described planning system in the previous paragraph.

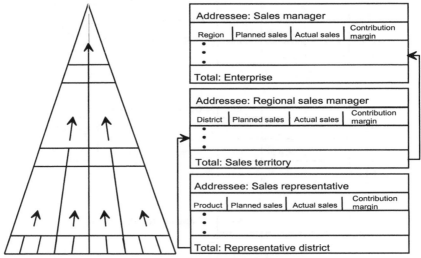

*Fig. 5.1.13/1    Aggregation or Compression Schema*

Figure 5.1.13/1 depicts the schema of an aggregation or condensation [Mertens/Griese 02, pp. 69-72]. Imagine that at the lowest level the sales representative receives a computer output (in print or on screen) in which one line appears for each product that he/she sold during the control period (e.g., during the last month). In that line planned sales, actual sales, marginal return, commissions and other information are provided. The row with summary data reflects also the sales figures he/she has achieved in that district covering all items/products he/she is selling. At the next higher aggregation level the rows with summary data of all sales representatives turn into the individual rows of the sales region. Correspondingly, on the next yet higher level of aggregation the rows with summary data for all sales regions become the single rows of a report intended for the sales managers of the enterprise.

## 5.2  Application Systems in the Services Sector

### 5.2.1  Particularities of the Services Sector

In the services sector, sometimes also referred to as the tertiary sector of the economy, when bundled with the information sector, more than half of all people are employed in the economy. This is generally the case for all industrialized nations and its contribution to the gross domestic product (GDP) increases steadily. Typical services sector firms are banks, insurance companies, trade, shipping, transportation, tourism and consulting companies. Moreover, one also considers the restaurant and hotel industry, self-employed professions, the entertainment and leisure time industries, education and health care, public administration or government, as well as particular forms of skilled labor, e.g., laundries, barber and beauty shops, or repair service shops to be part of this sector.

Contrasting services sector firms to industrial or generally manufacturing enterprises we may identify two major distinguishing characteristics:

■  The main component of the output, i.e. the service product, is immaterial.

■  During the production process a so-called external factor has to be involved. This is typically the customer him- or herself or an object within his/her possession, i.e. a tight relationship exists between the service provider and the customer.

In the provision of service products one may differentiate roughly among three phases from a business perspective:

1.  Implementation of the willingness to perform through a combination of internal production factors, e.g., making available means of transportation or an infrastructure for financial transactions.

2.  Performance agreement with the customer, e.g., the issuing of credit and insurance contracts or the booking of a trip.

3.  Service execution, e.g., the paying out of a credit, pay off of an insurance contract in case of an occurred damage or the actual flight to a travel destination.

Contingent on the immateriality of the actual service provision, the questions how the customer gains access to the offer and may utilize it directly are more important than warehousing and shipping problems. With performance agreements and the actual execution of delivery a direct and close customer contact is very often required. Service may be performed in the form of immaterial objects such as money and information, but may also manifest itself on a physical good, like the cleaning of a piece of laundry, the repair of a car or the shipping of freight. The performance of the service, i.e. the service product itself that the customer actually pays for always has an immaterial character. Often services are being sold bundled with physical goods, such as in a restaurant.

Information as an immaterial good takes on an important position as a production factor, aside from human job performance and operational resources such as the telephone and the PC. While in industrial manufacturing we are concerned with the planning and control of material flow, the production of service products often consists of the procurement, combination, processing and transmission of information, as well as the processing of documents serving as the carriers of information.

## 5.2.2 Information Systems Support of Service Processes

The possibilities to deploy targeted information systems in a service enterprise may be depicted in a phase schema of the service process (cf., fig. 5.2.2/1).

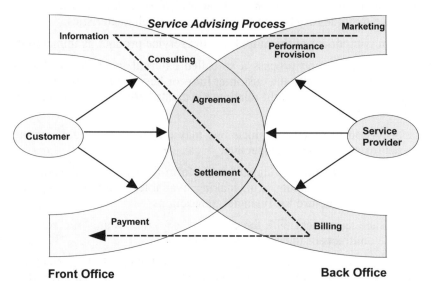

Fig. 5.2.2/1     Phases in the Service Delivery Process

Systems in support of the marketing and the information and advising phase make approaching the customer easier, from the communication of the benefits of the product on to the purchase decision. The phases of agreement and settlement are partly processed in contact with the customer, but in part also entirely within the organization. Again we may encounter special systems that facilitate agreement and settlement and make these phases more transparent and efficient.

Within the information systems architecture in support of process phases a delineation among systems distinguishes between the front office (having direct contact with the customer) and the back office (without direct customer contact). Furthermore systems may be structured in accordance to the enterprise's networked partners (cf., fig. 5.2.2/2). We may distinguish the following configurations:

- Internal networking within the service organization (among departments and employees)
- External networking with customers
- External networking with suppliers and partner organizations

In addition, we may encounter spanning networks such as in supply chain management (see section 5.4), as well as tools for specific tasks (e.g., procurement systems). The customer-oriented concepts and processes in the front office play a special role in the services sector of the economy. Due to the integration of the external factor into service delivery planning, configuration and provision, the increasing and often necessary individualization of the service product and other particularities (see section 5.2.1), electronic support at the customer interface is highly demanded. Sales-oriented approaches lead us into the area of electronic commerce (see section 5.3). When considering systems covering the entire customer life cycle we refer to those as customer relationship management (CRM, see section 5.2.3.1).

The customer gains access to the information systems of a service provider by establishing a connection at work or from home via a PC connected to a WAN (see section 2.5.3 and 2.5.4) or by using a self-service terminal. An important task of the *access system* is the checking of the customer's authenticity. Aside from the use of identification numbers and passwords this is accomplished increasingly through identification cards containing identification features and also a multitude of additional data. In the case of smartcards (see section 2.1.2) interactive checking mechanisms are possible as well.

Based on the immaterial character and frequent individualization of the delivered service we might address some additional thoughts to the topic of explaining the "final product". *Presentation systems* show the service product via multiple media in order to arouse the interest of the customer. Via enquiry systems it becomes possible to retrieve, evaluate and process offer and performance data, such as with stock market or travel information systems. Additional support is offered through *expert systems* that are designed to conduct an evaluation of the provided information (see section 4.3.2).

A particularity during the *negotiation* and *execution* of services (agreement and settlement phase) may be seen in the fact that the external factor is tied very interactively into the processes. In the front office area one often deploys transaction systems (see section 4.3.1.1). In advanced automated settings the customer may interact directly via a self-service, automated kiosk system or via a communication network with the information system of the service provider. In doing so he/she may configure an individual service preference and may also control the execution and accompany it in dialogue form. Additional support is offered by *software agents* that are capable to take on certain tasks and largely carry them out autonomously (see section 4.3.2.3). A customer may assign the search for offers on the Internet to a *search agent*. Also service providers may deploy *agent systems*, e.g., during an electronic auction on the Internet. When a potential buyer (e.g., a firm) may choose among offers by several service providers via an information and communication system and a supplier is enabled conversely to provide its products to several firms within a particular industry, we then refer to an *electronic market*. Examples are computer-supported securities exchange markets (see section 5.2.6.3) and travel sales systems.

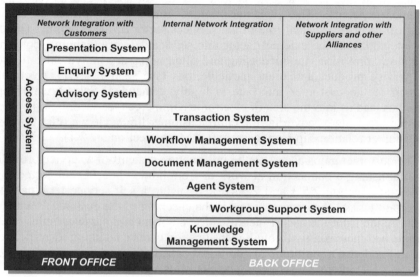

*Fig. 5.2.2/2    Front Office and Back Office Systems*

Frequently a full automation of service delivery does not make sense or cannot be done. In that case processes are often carried out within an *office environment* in the back office area of the service delivery organization. This is done through a division of labor, i.e. several employees cooperate during the delivery of a service product. Office information, communication and cooperation systems are deployed as information systems tools within a net-work of workplace computers and servers. Concrete support for the execution of processes is being offered by *transaction, workflow management* and *workgroup support systems* (see section 4.3.1). Thereby, one strives to achieve a paperless document flow, i.e. documents are collected, processed, deposited and transmitted in electronic business transaction folders. Documents are being archived, administered and made available via search mechanisms through *document management systems* (see section 4.3.1.3). Interfaces to the systems of customers and suppliers are very important, e.g., for acceptance, processing and transmission of applications through insurance brokers. If a direct, dedicated electronic connection exists to the application systems, it is possible to avoid down-times and the resulting losses in time, costs and quality. *Knowledge management systems* finally support the organization in the discovery, structure and availability of the resource "knowledge" (see section 4.3.1.5).

Just as in section 5.1 we are offering below various approaches in information systems support for service provision using examples. The phase schema in figure 5.2.2/1 will serve as our general guideline.

## 5.2.3  Marketing in the Service Process

### 5.2.3.1  Special Considerations

The role of Marketing is to determine the product, price, distribution and communication policy. During the service delivery process the marketing phase serves specifically the task of contacting customers with the goal of a purchase of the service product. It comprises the "one-to-many" form of communication in which especially the service provider becomes active. During the continued course of the service process, i.e. in the information and advising phase, the initiative originates with the customer who reacts to the information offer by the service provider. This step, however, typically occurs in the "one-to-one" form of communication. The soliciting of services is made more difficult by the immaterial nature of the product and the often individual delivery. In contrast to completed and presentable material products, the precise demonstration of utility of services to potential customers prior to their purchase is difficult.

In the decision which services the organization should offer, a first step is an examination of the demand side. Market research activities, as well as subsequent descriptions of the markets or market segments focus on the demonstration of particular features with regard to characteristics pertaining to customer demand and needs. The often required direct contact to the customer, e.g., offers the possibility to inquire issues pertaining to customer satisfaction directly with regard to the already provided service delivery. Moreover, we have the opportunity to establish a customer profile that takes into account, e.g., the special characteristics of the external factor, the temporal distribution of demand, as well as the desired intensity of the willingness to perform.

A comprehensive concept for managing the relation to the customer is the *customer relationship management* approach (see section 5.1.2.4). A basis for this is the networking of various application systems in the service provider organization through which customer data may be captured and stored. Through the use of data mining it is possible to discover, e.g., preferences of customers of a travel agency. The aim is to design service products tuned to their individual needs (e.g., adventure travel). Efforts addressing the individual customer are especially emphasized. In particular large service companies such as banks, insurance companies and mail order firms attempt to increase the relationship to the customer and to assure it for the long-term in this fashion.

### 5.2.3.2    Use of the Internet

An important element of the organization's communication policy is the use of the Internet. In *one-to-many communication* classical marketing concepts can be applied. An organization's own website, as well as the use of websites of other providers, enable various forms of marketing communication.

Application areas for e-mail are especially asynchronous communication, e.g., in the service area or when handling customer queries or complaints, as well as for sales advertising, e.g., in the form of electronic advertising letters. Newsletters appear regularly for a fixed circle of subscribers and contain new product or price information.

Banners are advertisements or placards on the WWW. They are mostly animated graphics with a link to a firm's own website. They announce the own sites, enable the ordering of information material or may offer small games and prizes. The possibility to have one's own banner appear on the visitor's site when certain keywords are entered is intriguing. For example, when entering the words "car" or "Mercedes" into a search engine we will find also a banner advertisement that advertises for Mercedes products (maybe by the car's manufacturer, a Mercedes car dealer, etc.) on the results page. So-called rich-media banners offer multimedia depiction and also interaction possibilities linked with the banner itself (e.g., the ordering of products).

Various forms of web pages are presented in figure 5.2.3.2/1. While electronic business cards usually don't exceed one or two pages and mainly serve public relations of the firm, electronic prospectuses and catalogues present the product palette. Such catalogues come in many different shapes and forms depending on whether the products are merely to be depicted and listed or whether certain functionalities are to be added such as customer service, product configuration, and testing. So-called "advertainment" sites offer advertisement together with entertainment, e.g., games. Online shops serve advertising and sales within the framework of electronic commerce (see section 5.3). Virtual communities are theme-related meeting places on the Internet. The community among its members emerges through emotional, motivational and cognitive relations (e.g., the same interests, needs, knowledge). From a marketing perspective, this constitutes a contact potential and one assumes that members of a community also have similar preferences and demand similar products. Through various self-reinforcement effects better sales offers, a dedicated customer base and finally a dedicated clientele evolve that will be embraced by customer relationship management.

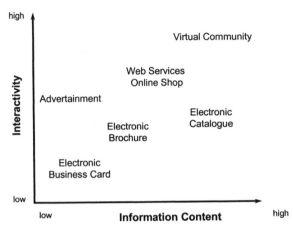

*Fig. 5.2.3.2/1    Types of Websites*

### 5.2.3.3    Micro-geographic Systems in Insurance Marketing

The classical approach for differentiation strategies is *market segmentation*. One considers simultaneously a multitude of segmenting characteristics and develops fine-tuned, multidimensional consumer typologies and describes them in great detail. In turn, these then permit the derivation of attractive target groups (e.g., TAPS = technically advanced persons). Problematic, however, is that these groups cannot be localized in the market for targeted communication. At this weak point computer-supported micro-geographic market segmentation is utilized in which, instead of viewing individuals, tightly delimited, finely-tuned spatial residential areas are examined. This method is built on the assumptions of *segregation* and the *neighborhood effect*, respectively. The latter denotes that people with similar life styles and similar consumption habits settle in spatial neighborhoods. Through the demarcation of these residential area groups of sufficiently homogeneous persons with comparable demand structures may be determined.

Basis for a *micro-geographic system* is the design of an area-wide structure that subdivides the regional total market into finely-tuned geographic parcels such as voting districts, street sections or building clusters (cf., fig. 5.2.3.3/1). By linking this *parcel database* with *marketing databases* from various sources a *regional database* is being developed with one data set per parcel. Each parcel then features about seven to 4,500 households. This essentially constitutes the core of a micro-geographic system. Aside from geographic information the parcel data contain characteristics such as age, education and income structure of the households. Similar market parcels are identifiable through data pattern recognition processes yielding clusters that are being referred to as *geotypes*. All parcels within a geotype feature similar structures and permit the assumption of a similar consumption behavior.

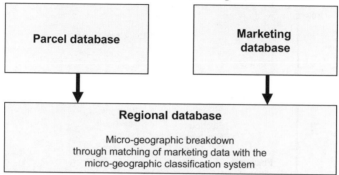

*Fig. 5.2.3.3/1   Micro-geographic Database*

For example, an insurance company may conduct an evaluation of geotypes based on internally available statistics in order to build marketing management programs based on this knowledge. If a geotype shows few contracts but high insurance payouts, it may be recommended to stop courting this segment or to adjust the rating policy to the level of prevailing claims and losses. Another application is the targeted selection of sales representatives based on the geotype profile, especially for attractive customer segments.

Beyond that an insurance firm may develop an MIS (see section 4.3.2.2) for regional marketing purposes in order to support the sales manager to localize targeted high risk and chancy sales areas or strengths and weaknesses in the marketing management effort.

### 5.2.3.4    Sales Force Support in Insurance Sales

When acquiring insurance contracts personal contacts are of great importance. The possibilities for self-service solutions in the insurance business have been developing considerably slower than in the banking area. Aside from the increasing use of the Internet for direct sales of insurance products the electronic integration of the insurance central office, field service and customers leads to a transformation of the sales organization. The customer may request offers, e.g., for car and home insurance, or he/she may agree to an informational meeting. The customer may receive the answer to the query optionally by traditional or by *electronic mail*. Central to information systems support in this setting is the requirement of flexible access to the database of the insurance firm or to calculate insurance service variations on the spot. This is possible through the use of notebooks and public communication services. A typical process may look like this:

■ The sales representative evaluates the stored customer profiles and recognizes that a customer has a life insurance contract for a relatively low sum of money and does not have any work disability insurance. In addition the salesperson recognizes that the customer is a professional truck driver, who is married and recently became a father.

■ The employee creates an individual letter interactively using text boiler plates (in which the researched data are integrated) addressed to the customer and asks for an appointment.

■ The computer manages the agreed upon and planned dates and issues, reminders and follow-up messages.

■ When visiting the customer the notebook is used to calculate various alternatives to increase the insurance policy's coverage and variations for a disability insurance are presented.

■ During the course of exploring various insurance policy coverages and options the needs for the customer may be identified more precisely so that the system may advance a detailed proposal and solution.

■ When accepting the proposal the application forms for the revision of the life insurance policy and the new disability coverage are printed out.

■ After the contracts are signed by the customer the sales representative transmits the data via mobile phone to the central computer so that the application may be checked.

■ If this review of the applications can be done immediately, the acceptance confirmations may be transmitted back to the sales representative right away. Then the customer will receive the insurance policies after they are printed out on the printer connected to the notebook.

Usually the most important insurance rates, as well as the district's customer data will be available on the notebook. The common questions can be answered with the help of appropriate program packages for advising, rate calculations and application processing. The customer data may be augmented by additional profile information which may include "soft" data pertaining to the customer's personal and financial situation which the sales representative acquired in personal meetings.

Aside from operational tasks application systems take on more and more *advising tasks* similar to banking. Accordingly, we find, e.g., expert systems (see section 4.3.2.3) for the support of sales representatives in areas where he/she may not be a specialist, as well as in support of the specialist examining and evaluating damage claims.

## 5.2.4  Performance Provision in the Service Process

### 5.2.4.1  Special Considerations

One may distinguish between the performance potential, i.e. the production factors available in the organization, such as human and operating resources, and the current performance provision for production. Both must be planned particularly precisely in the service organization, due to the necessary external factor and the frequent "perishableness" of the service product. This perishableness arises from the immaterial character of the service. The available performance potential often expires through disuse of the service, since the usually time-dependent service offer is not storable.

During the actual service process the organization-internal factors are combined in advance as much as possible without considering the external factor. As a result a service provision benefit is generated, i.e. the service is available for a potential customer.

When considering an organization's willingness to meet the today uncertain future demand for its services we may consider various approaches:

- Total demand is estimated (e.g., from sales planning or via time series analyses) and the willingness to perform is aligned accordingly (deterministic procedure).

- With the demand for material, e.g., bandage material in the hospital, the willingness to perform may be realized differently. When a reorder level (see section 5.1.3) has been reached, the used portion of the inventory is replenished by placing a procurement order (consumption-driven procedure).

- As part of product differentiation differently usable performance potentials lead to different forms of fulfilment of the same basic product. Considering an airline ticket in the better-priced economy class vs. the more expensive business class is a fitting example. In this case the total capacity has to be divided in accordance with predicted individual demands.

- Aside from its willingness to perform an organization may also influence demand (e.g., through pricing) to a certain degree.

### 5.2.4.2  Yield-Management Systems in the Tourism Industry

During quantitative order planning yield-management systems (YMS) may support optimal capacity planning and may avoid unused capacity. Service products are often perishable products. An empty seat on an airplane that has just started the engines can no longer be sold. YMS are utilized by airlines, car rental firms, hospitals, hotels, etc. American Airlines'™ SABRE system, e.g., unintentionally "discovered" this yield management potential after having designed the system initially mainly as a reservation system.

YMS serve the maximization of yield by adapting capacities in different classes to actual demand and at the same time attempt to control demand behavior by applying certain pricing strategies. In doing so the following strategies need to be considered among others (cf., fig. 5.2.4.2/1):

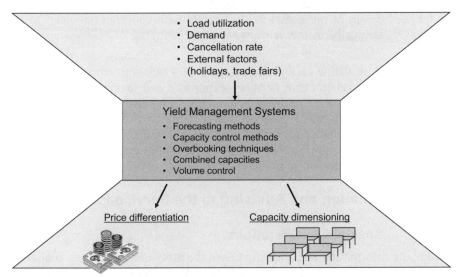

*Fig. 5.2.4.2/1    Yield-Management Systems*

Prices may be differentiated based on differing criteria:

■ Temporal price differentiation (price variation): The product has different prices at different times.

■ Quantitative price differentiation: High-volume customers may take advantage of more favorable prices.

■ Market-segmentation price differentiation: Through market segmentation (see section 5.2.3.2) demand groups may be identified showing different characteristics and preferences. These groups are offered different prices and possibly also different product variations.

Fine-tuned prediction methods are the basis of all YMS. Based on known parameters and by the estimation of unknown parameters the system forecasts the demand to be expected in various product categories (e.g., booking classes) for a particular point in time (stochastic component) on the one hand. On the other hand, one also derives the reservation and cancellation process from historical data over time (dynamic component). In that way we may compare the forecast reservation process with the actual one such that timely corrections may be made if there are deviations.

An additional technique of YMS is to make deliberate and targeted overbooking recommendations based on cancellation history. This is especially possible when high-priced offers are being sold after low-priced offers. As a consequence, an upgrading (e.g., better hotel) of the lower-priced product can take place if the rate of cancellations turns out to be smaller than its forecast. Nevertheless a comprehensive upgrading has a negative effect on the overall receipts of the service provider since the more expensive substitute offers can no longer be sold at the actual price. Trends with cancellations need to be watched dynamically in the same manner as booking or reservation processes.

In the case of airline companies the goal of YMS is to calculate a profit-maximized offer relationship between expensive and favorably-priced seats by using capacity control methods. A passenger who cannot be provided with a seat in a more expensive class implies a larger loss than an unaccommodated passenger in a more favorably-priced booking class. The particular difficulty here is based on the fact that more favorably-priced bookings occur temporally prior to the full-priced bookings.

## 5.2.5  Information and Advising in the Service Process

### 5.2.5.1    Special Considerations

During the information and advising phase the problems arising are similar to these in advertising. The product is immaterial, the service delivery is often individual and communicating the benefit is difficult. The attention of the customer is already aroused in this phase, yet two more decisions will have to be made. One is basically the actual acquisition of the service, the other is the choice among different variations.

Information in this phase should be as current as possible; access should be simple and should not take much time. In addition the presentation should convey trust to the interested party and one should describe the desired service as well as possible.

During advising the service provider organization helps the customer, e.g., in the selection among various choices. In doing so, special customer needs have to be considered. Since the advising often occurs through direct contact to the customer, time restrictions also play a role. The quality of advising thus expresses itself not only in the results, but also in the time frame needed to achieve these results.

### 5.2.5.2    Enquiry Systems in Passenger Traffic

Many organizations in the passenger carrier service provide an entire range of offers to private or business customers, from rate schedules up to actual booking via the Internet.

Often the customer is able to retrieve recorded information about arrival and departure times. Those voice-based services through which access to data is possible by using the telephone, are referred to as *audiotex*. Bookings and credit card numbers are being stored by the digitized recording of voice entries. Dialog control is possible for each voice entry or via the caller's telephone keypad.

Aside from the original information terminals in the airline carrier business, merely offering departure time, destination, last-minute flights, etc., more recently we may observe the use of automated ticket machines where the customer, who has already made and paid for a reservation, may pick up the airline ticket. Additional automation efforts are to enable the identification of the customer and payment with credit cards. At the same time an automatic check-in is to occur. In the very near future the issueing of a currently paper-based passenger ticket will belong to the past, i.e. all tickets issued will be electronic or e-tickets.

*PRACTICAL EXAMPLES*

*The electronic time schedule of the German railway system (Deutsche Bahn™ AG) comprises connections among 50,000 train stations across Europe and more than 180,000 stops for the public passenger traffic for its bus system in Germany. The data contain the complete German local and long-distance traffic operated by Deutsche Bahn AG, the train connections to all larger European cities and also the time schedules of individual European countries. Aside from this detailed time schedule information the system also offers fare information. Subsequently the customer may wish to order a ticket and a seat reservation electronically. Payment takes place by using one's credit card or direct debit authorization.*

*Aside from railway companies, most airlines such as Lufthansa™ AG are offering access to their flight schedules as well. The provided information shows the availability of all connections and additional flight information (travel time, etc.). The flight reservation may be made directly on the website and, subsequently, the Lufthansa central reservation system registers these bookings. The reservation is then confirmed by sending an email to the customer.*

## 5.2.5.3 Advisory Systems in Retail Sales

Advisory systems are an attractive means of automating service processes in retail sales. They are to facilitate choosing among available offers in need of an explanation and making the right choice. Moreover, they are to unburden the salespeople from their advising task. The extreme setting here is "*self-advising*" at a *point-of-information* (POI), as a feature of self-service applications. For demanding advising tasks we may want to consider expert systems (see section 4.3.2.3).

*PRACTICAL EXAMPLES*

*In various kitchen stores a dialogue system is being used in which the wishes and certain conditions (e.g., measurements of the kitchen, position of windows, direction of the door opening) of the customer are entered. The application system places the chosen kitchen furniture and devices in the available space. By using a mouse the customer may move the furniture on the monitor. Finally the machine shows the customer a three-dimensional depiction so that he/she gets a pretty good impression of the future kitchen. At the same time the cost of the chosen kitchen is calculated. All of this may be printed out on paper so that the customer receives a drawing and layout, as well as a listing of prices for the items and service.*

*With the help of software for advising customers on the choice of eye glasses the sales person assigns the customer a particular type (sporty, elegant, etc.) and elicits requests (e.g., light conditions, using a monitor at one's workplace). From this information the application system derives certain conclusions about the respective glass characteristics (e.g., breaking strength, weight, potential glare, tint, price). The advisory system delivers a ranking of the alternative glasses, suggests the optimal allowance and calculates the sales price for each combination. Beyond this the optometrist has the possibility to request a detailed description of the recommended glasses. In case the customer is not happy with the suggestions, he/she may modify the choices.*

### 5.2.5.4    Advisory Systems in the Bank Investments Area

Special AS support the investments advisor in the selection of securities for the customer, as well as the analyst during his research on the forecasting of market trends.

Important tasks of the software are to be assigned to the following areas (cf., fig. 5.2.5.4/1):

- Support of the investment advisor
  - Analysis of stock prices and end-of-year statements of the companies to be considered
  - Analysis of the customer's investments in order to make inferences about the investment profile
  - Individual asset allocation recommendations through the use of a filter system
- Support of the analyst
  - Analysis of the relationships among individual stocks
  - Forecast of the enterprise, stock and market developments

Aside from expert systems more recently other knowledge-based techniques (see section 4.3.2.3) are deployed for decision making, e.g., artificial neural networks for stock price forecasts.

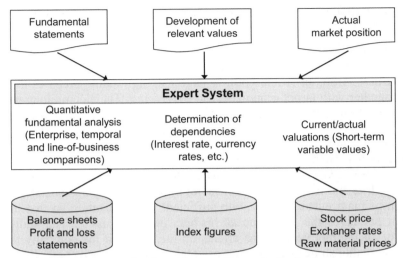

*Fig. 5.2.5.4/1    An Example of an Advisory System in the Investment Area*

## 5.2.6  Agreement in the Service Process

### 5.2.6.1  Special Considerations

As a consequence of advertising or the information and advising phase, respectively, the customer knows the services of a service provider or the offers of several firms. Offers include the description of the service performance that the provider wants to sell, as well as the corresponding price.

In the most simple case the general service offer may be accepted by the customer without any changes. After a binding agreement (acceptance of the offer) has been reached, one quickly progresses to a performance and payment promise, resulting in a contract. Often offers have to be adapted individually. After a performance specification the service provider makes an individual offer that the customer in turn may accept immediately. In addition it is also possible that customers advance own suggestions in order to change the price/performance relationship in their favor. They may then make a purchase counter offer or negotiate prices, discounts and conditions. When selling services such as flights via the Internet, various forms of auctions are gaining increasing importance, including reverse auctions seeking the cheapest supplier.

### 5.2.6.2  Individualized Magazine Offers

The dispersion of information services via electronic networks simplifies the individualization of services greatly. So, some magazines of online subscribers offer the possibility to receive a so-called *personal journal* that includes only the desired themes identified by the reader and is edited accordingly (cf. fig. 5.2.6.2/1).

To accomplish a personal journal a reader profile is established and a selection module using the developed classification scheme assembles the desired articles from an article pool. A page-making module prepares the articles, together with target group-specific advertising. The electronic magazine has connections via hyperlinks (cross references to other documents, see section 2.2.2.1.1) to the complete article pool and to additional information such as maps. These may be prepared by the editorial office and might originate from external databases. It appears then to the customer that a "personal issue" of the magazine was created just for him or her. This, however, is only partially true and in an absolute sense not a form of true personalization. It is a form of mass customization with sufficient personalization so that the recipient perceives the magazine as his/her "personal" issue. The additional offer of an Internet version has effects on the image of the company that in doing so demonstrates its customer orientation and innovativeness. It is often possible that the potential customer receives a sample issue of the individualized product.

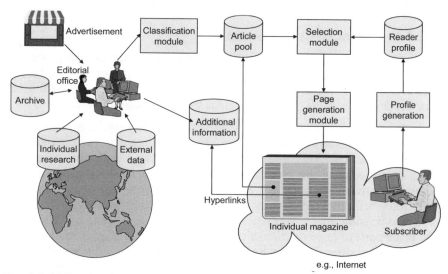

*Fig. 5.2.6.2/1    Production of an Individualized Magazine*

### 5.2.6.3    Securities Trading at the Electronic Stock Exchange

With homogeneous goods in transparent markets more and more electronic systems are being used, primarily to coordinate demand and supply, to make price formation possible and to complete contracts e.g., by an auction. Often these systems are components of *electronic markets* that link the systems of suppliers and buyers. An example of this is the electronic trading system

*Xetra™* (Exchange Electronic Trading) introduced in 1997 by the German Stock Exchange in Frankfurt (Deutsche Börse™ AG). With it all listed stock values, stock options and public bonds are electronically tradable. The system's core function is the support of the agreement phase. Xetra is based on a client-server architecture (see section 2.4.4) in which the participant installation (Xetra Front End) with the stock traders consists of one or more participant servers and of workstations (dealer places). The central stock exchange functions are carried out from the Xetra Back End server (cf., fig. 5.2.6.3/1).

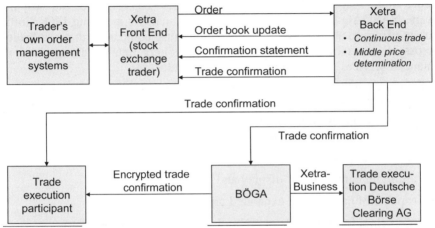

*Fig. 5.2.6.3/1    Xetra Trade Execution*

The market model implemented within Xetra supports pricing, including the middle price determination, as well as continuous trade. During the middle price determination the stock price at which most business transactions may be completed is discovered once while considering the entire order situation. For continuous trade the system has an order book in which buying and selling orders are matched and settled immediately. The order book is open to all participants so that market transparency is warranted.

A Xetra participant may enter his/her own orders over the Xetra Front End or over an Xetra-connected private system. Moreover, the participant may observe the overall order situation via an order book update function. If an order transaction is completed, the Xetra Back End generates and transmits a confirmation of order execution. For documentary reasons and in order to initiate stock exchange and money transfer the Xetra Back End transmits a corresponding trade confirmation to all participants. Every transaction obtains a unique transaction number or identifier so that the status of an order is comprehensible any time.

The Deutsche Börse Clearing AG provides the central clearing and settlement system available for the completed businesses in Xetra and makes the funds available for the necessary security and fund transfers. The interface with Xetra is the BÖGA (Börsengeschäftsabwicklung or stock exchange transactions) system that transfers the data of Xetra transactions automatically.

A somewhat parallel development was observable on the NASDAQ Stock Market, Inc. (NASDAQ) with the roll-out of a new software entitled Super-Montage, a fully integrated order display and execution system for the trading of NASDAQ-listed securities. The juxtaposition of Xetra and SuperMontage is an interesting comparison. Listing nearly 4,000 companies, NASDAQ trades more shares per day than any other U.S. market. Over the past five years, NASDAQ has outpaced all other U.S. markets in listing IPOs (initial public offering). It is home to category-defining companies that are leaders across all areas of business including technology, retail, communications, financial services, media and biotechnology industries.

NASDAQ takes in data from more than 350,000 terminals from around the world and processes more than 5,000 transactions per second. SuperMontage's new computer architecture allows greater speed and efficiency to the market and also supports higher order quantities.

SuperMontage is:

- a fully integrated order display and execution system,

- capable of handling an expanded universe of orders, and

- in its final form, completely rebuilt on a scalable infrastructure.

## 5.2.7   Settlement in the Service Process

### 5.2.7.1   Special Considerations

The range of services in the various industries and enterprises of the tertiary sector is very multifaceted. An attempt to classify the various types of services is based on the differentiation between performance object and performance impact (cf., fig. 5.2.7.1/1).

In the definition of the service production the possible performance objects are:

- persons
- physical goods
- information
- nominal goods (e.g., money)

The performance impacts are related to:

- the material or the object itself
- the space, i.e. the location where the performance object is located
- the time during which the performance object is available

This classification is a relatively rough approach. Services may often not only be assigned to one category, but they may also comprise several performance objects, as well as performance impacts.

| Performance impact / Performance object | Object-changing | Distance-spanning | Time-spanning |
|---|---|---|---|
| Persons | Health care services | Passenger traffic | Hotel business, entertainment service |
| Physical goods | Repair shops | Transportation of cargo, trade | Warehousing |
| Information | Government, publishers | Telecommunication | Libraries |
| Nominal goods, e.g., money | Banks | Monetary transactions | Insurance industry, real estate firms |

*Fig. 5.2.7.1/1    Classification of Services with Examples*

In the following we will introduce two from the multitude of information system applications that are object-changing and in each case one from the space-spanning or distance-spanning, as well as one from the time-spanning category.

Considering the integration of the customer when carrying out the service, we may distinguish among three cases:

■ The service is only carried out in the Front Office: This may be the case when little planning and support services are required, e.g., as with some manual work or craftsmen services.

■ The service is carried out in the Back Office exclusively: For this the customer must have specified a contract during the agreement phase that gets completed without any direct involvement of the customer. The performance object is often information or nominal goods, such as securities transactions that a bank conducts on behalf of the customer.

■ The service is being provided partially in the Front Office and partially in the Back Office. This is especially important when we are dealing with services in which the customer constitutes the performance object. Here one encounters a tight linkage between the Front and Back Office, e.g., in the hotel business and the field of medicine (see sections 5.2.7.5 and 5.2.7.2). Accordingly, drawing a sample of blood will occur in the Front Office, whereas its analysis in the laboratory will take place in the Back Office.

## 5.2.7.2    Services in the Health Care Sector

Diagnostics systems support the user on the basis of necessary, sufficient or exclusionary symptoms or findings, respectively, in the decision about the illness to be diagnosed. In addition, more sophisticated systems quantify the confidence of the diagnosis, as well as the expected diagnostic benefit of further examinations. In this area expert systems are often deployed (see section 4.3.2.3) .

Here one distinguishes:

- *Screening systems* that serve the systematic acquisition of patient data with the goal to find indicators of illnesses or predispositions

- *Laboratory diagnostic systems* that support the interpretation of laboratory results

- *Consultation systems* that either confirm a diagnosis ordered by the physician or suggest other diagnostic or therapeutic possibilities

- *Electronic textbooks* within the field of medicine

- *Patient management systems* that, e.g., support decisions about a return visit for a follow-up examination or the date for an operation

In the surgery and intensive care areas monitoring systems are deployed that continually control the captured medical data streams.

In advanced solutions the physician may retrieve all relevant patient data during a patient round using a pocket computer, i.e. PDA. By tapping appropriate text boiler plates the documentation for diagnostic findings are generated directly at the patient's bedside. All newly captured data are entered into the electronic patient record.

Information systems are also used in the calculation of diet plans and the dosage of medications and infusions.

In the medical office information system components are used which assist the physician in the documentation effort and in his/her correspondence. They are also providing the physician with information.

Important documents are:

- *Medical history reports (history of an illness)*
- *Reports on diagnostic findings*
- *Physician letters*
- *Service and performance listings, as well as medical documentation*

Information systems mainly support the nursing staff with all administrative tasks. As part of the medical documentation effort we observe increasingly a care documentation in which all patient-related information from patient rounds documentation, shift change documentation, drug prescription plans, etc. is merging so that the nursing staff may draw upon all information from similar cases of the past. An additional typical application area for information systems is the creation of staffing, bed occupation and operating room usage plans for the station.

In the administration of service and performance delivery the following system components take on a prominent role:

- *Patient administration*
  Patient in-take (inpatient and outpatient), capture and care of patient master records, occupancy, relocation and discharge documentation; creation of statistics
- *Patient billing/accounting*
  Processing of cost absorption applications, account settlement of in- and outpatients, writing of invoices for health plans and individual patients, consideration of lump-sum allowances per case
- *Recording of services rendered*
  Recording of services in the various functional units in the hospital and institutes, certification of required service and performance delivery, as well as statistical evaluation
- *Cost accounting*
  Recording of occurred costs, assigning of certified service and performance delivery to cost centers, creation of the cost accounting report

Beyond these additional business administrative functions need to be supported, such as financial accounting, human resources, asset and material management.

Similar to hotel management, hospitals run stock-keeping and warehousing programs and systems for material usage plans, controls and documentation. Examples are food for feeding the patients or medication, bandage material, blood supplies, etc. within the hospital's pharmacy.

Customer (here: patient) data are needed in many different scopes of duty, e.g., in feeding the patients, clinical research, as well as with administrative tasks. A hospital information system has the task to make this information available (cf. fig. 5.2.7.2/1). In addition, often external information has to be acquired in order to complete the service or performance delivery.

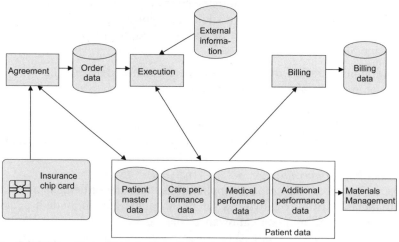

*Fig. 5.2.7.2/1    Data Flow in Hospitals*

By networking hospitals (including university hospitals), established physicians (general practitioners and specialists) and laboratories, distributed specialized knowledge and diagnostic findings (e.g., X-ray pictures) may be exchanged. Moreover, it is possible to integrate special services, such as a centralized picture archive or picture development systems (cf., fig. 5.2.7.2/2).

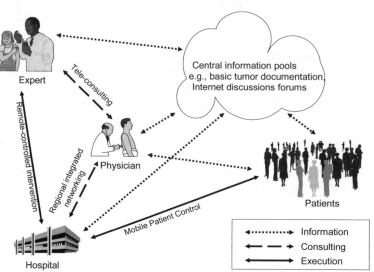

*Fig. 5.2.7.2/2    Distributed Applications in the Medical Field*

Some advantages are:

- General practitioners in remote areas may improve the quality of their local services via the use of tele-consultation by experts in the particular field, reduce the risk of wrong decisions, and reduce the additional burden for the patient having to travel to see the expert.

- Multiple examinations and patient transfers may be avoided.

- In emergencies the networked settings may lead to considerable time savings when acquiring findings and diagnostic results. Many hospitals today are linked and stay in contact with university hospitals and others by various telecommunication means via a bi-directional data, video and voice connection in real time. Consequently, this implies a considerable improvement (quality, accuracy, speed and time) in local emergency medical care.

- Using techniques developed for the remote control of robots in space, it has become possible for a surgeon to carry out an operation on a patient even though the two are geographically distant to each other. This example shows again that often today distance no longer plays a major role when working with information systems.

- With mobile patient monitoring the distance between physician and patient is bridged. The patient or his/her bio-signals may be monitored from the patient's home. For this one uses, e.g., miniaturized bio-signal amplifiers or measurement devices for the pulse rate, etc. The transmission of the data may occur, e.g., over mobile phone networks. Aside from the advantage of avoiding longer hospital stays or advanced dismissal of a patient after an operation, observations within everyday life may be revealing by itself.

### 5.2.7.3 Awarding Bank Credit

The handling of credit applications with banks is usually a well-structured process that is carried out by following defined rules. Therefore, *Workflow Management Systems* (WMS) are frequently used to perform this service task (cf., fig. 5.2.7.3/1).

The WMS starts with an information and advising module. If the customer submits a written application, it is captured by the document management system. Based on the application the credit analysis starts. For this the WMS opens the process folder and makes the internal credit record (data from the credit user database about all credits and collaterals) available. In the next step the WMS creates a credit protocol that in summary form contains the situation of the applicant while considering the present application for credit. Additionally it initiates the inquiry of the external credit record, obtained from usually one of the three credit bureaux in the United States: Equifax, Experian and Transunion. Each person within the credit department's systems is given a credit score based on his/her past and current performance and risk level.

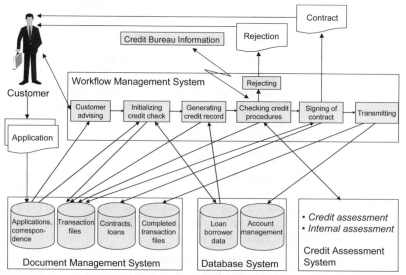

*Fig. 5.2.7.3/1    Generalized Concept of WMS in the Credit Application Process*

Credit scoring is a system creditors use to help determine a person's credit-worthiness. Information about a person and his/her credit experiences, such as bill-paying history, the number and type of active and dormant accounts, late payments, collection actions, outstanding debts, and the age of the accounts, is collected from the credit application and the applicant's credit report. Using a statistical program, creditors compare this information to the credit performance of consumers with similar profiles. A credit scoring system awards points for each factor that helps to predict who is most likely to repay a debt. A total number of points, i.e. a credit score, helps to predict the credit-worthiness of the applicant, i.e. how likely it is that the applicant will repay a loan and make payments when they are due. Apart from this one uses various statistical techniques or e.g. artificial neural networks for the credit checking.

After that the credit protocol is transmitted to the credit checking system and the WMS deposits it into a process folder. If the credit is backed by the bank, the agreed upon amount may be transferred into the customer's bank account. If the application was denied, an appropriate notification will be mailed to the customer. In the end, the entire process folder will be archived.

## 5.2.7.4    Freight Forwarding

In the business of freight forwarding dramatic changes have occurred as a consequence of modern information technology developments. More and more traditional shipping companies are growing into the role of logistics providers. In the following example [Mertens 01, pp. 103-104] the Rhenus AG takes on assembly functions aside from the classical logistics functions of warehousing and shipping. In doing so we move a little bit into the direction of supply chain management (SCM) (see section 5.4).

*A PRACTICAL EXAMPLE*

*The Dynamit Nobel Kunststoff GmbH (DN) in Weissenburg, Germany, a plastics manufacturing company, produces lacquered plastic bumper modules for the automobile industry. Audi built together with Rhenus AG a so-called Systems Center in the city of Heilbronn. Here, the different versions are being assembled using the appropriate bumper components.*

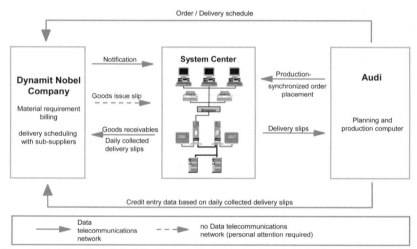

**Data communication**

Fig. 5.2.7.4/1    *Simplified Technical Connection between the Dynamit Nobel Company, its Systems Center and the Customer*

*Figure 5.2.7.4/1 depicts in simplified form the technical connection among the three partners. The customer transmits the delivery schedules (weekly) and the fine-tuned delivery schedules (daily) via the data telecommunications network to Dynamit Nobel (DN). For the shipped goods the corresponding invoices and credit memos are exchanged in the same way. The Systems Center receives from the customer the sequenced delivery schedules (production-synchronous schedules), whereupon the corresponding assembly steps are completed and the bumpers are delivered to the customer. The note of delivery is also transmitted via the network by the Systems Center. An intensive data exchange occurs also between the Systems Center and the DN. Each shipment from DN to the Systems Center is announced electronically. After the goods have been delivered to the Systems Center a goods received confirmation is sent. Moreover, a cumulative delivery note is transmitted to DN on a daily basis. Merely the waybill is still physically exchanged between the supplier and the Systems Center [Krill 96].*

## 5.2.7.5    Services in the Hotel Business

When making a hotel reservation the employee in charge will have to retrieve data about actual occupancy and current, as well as future availability. After the customer's preferences have been entered, the system suggests available rooms. The status of the chosen room is then changed to "reserved" for the desired time period. A reservation confirmation and the form for the intake of the customer data at actual arrival is automatically printed. An alternative for this procedure is the reservation through a central computer-supported reservation system that, e.g., is accessible via travel agencies or via the Internet (cf., fig. 5.2.7.5/1).

When the guest arrives in the hotel the system creates a guest account where all debits and credits incurred by the customer during his stay may be charged. The room status changes to "occupied". Any master data about the guest that are still missing are being transferred by a program from the registration form into the master data file or they are being asked off the monitor and are then entered. Moreover, the telephone and video equipment is made available automatically.

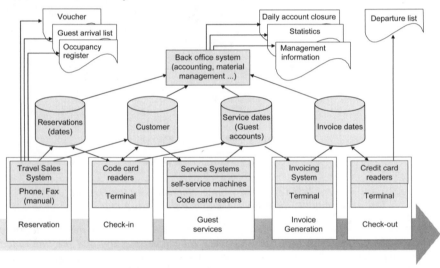

*Fig. 5.2.7.5/1    Processes in the Hotel Business*

The utilization and recording of services may be greatly simplified through hard- and software components. Code cards and keys not only increase the level of security, but they are also machine-readable and can be used as internal credit and booking cards.

Routine data in the restaurant and bar areas are transmitted via computer-controlled cash registers. An information and entertainment program using TV and video communication equipment (current guest account balance, electronic mail, etc.) complements the computer-supported customer service.

An integrated system ensures through online connections to subsystems like restaurant cash registers, telephone equipment, etc. that the guest account is always up-to-date. Consequently, a current invoice may be issued at any time. The check-out process is thus greatly facilitated and sped up; the data flow immediately into the accounting system. The telephone and television systems are automatically closed as soon as the check-out occurs and the room status is reset to "available".

## 5.2.8 Billing and Payment in the Service Process

### 5.2.8.1 Special Considerations

A problem with invoicing for services is their high degree of individuality. Often this requires the determination of an individual price for each customer (see YMS, section 5.2.4.2). In addition, due to the immaterial character of the service, it is not always easy to price partial service in a detailed fashion.

The invoicing problem may be solved in two ways:

- Previous agreement on a fixed price for the entire service, if the scope of services, e.g., is predetermined through some form of standardization (*product-related approach*)

- Final itemization of a comprehensive invoice that is derived from the effort of the service provider by gathering the respective components of the service and by summing up the individual prices (*process-related approach*)

An advantage of the product-related approach is that the customer already knows which costs are to be expected before the service is delivered. When the services are to be delivered to individuals, it is possible to check on the entitlement for the service very easily, e.g., by issuing an admissions ticket that has to be acquired in advance. This principle may be used mainly when mass service deliveries with a known and specifiable effort or a lower level of individuality, like public transportation, are to be provided.

When issuing the invoice after the service delivery the individual components of the service are gathered so that the service provider is running a lower risk of not covering the expenses with the revenue. Admittedly, the final invoice may only be issued after the completion of the service delivery. The delivered services and performances need to be stated explicitly and comprehensively for each respective customer.

Since direct contact with the customer is often required and since there are no sales intermediaries, service-oriented companies have to take care of many more individual payment processes than businesses that manufacture material goods (often sold through the trade professions (intermediaries) to the final customer). The classical means of payment 'cash' not only harbors a degree of inconvenience but especially the risk of forgery, loss or theft (quality side), as well as considerable efforts that are needed for safe-keeping or transport to a bank (productivity side). When the physical meeting of buyer and seller may be avoided through the use of self-service concepts or telecommunication media, it then behooves us to carry out the payment process over the same medium (e.g., the Internet).

The following payment methods demonstrate the tight connection of service producers, supporting financial service providers and customers in this phase of the service delivery process.

### 5.2.8.2    Point-of-Sale Billing and Payment

Assume that you are shopping in a supermarket and you pay at the cash register, i.e. the point where the actual purchase occurs legally (actually a verbal contract to buy and sell something has occurred). This point or location is referred to as *Point of Sale* (POS). The cash register system (POS system) reads the bar code affixed to the product that is standardized regionally world-wide. Figure 5.2.8.2/1 shows the North American UPC and the European EAN bar codes; each is described briefly.

This data capture process triggers a good number of sequential events in a mature, integrated information system:

1. Intra-enterprise
    a) The system searches and retrieves the item description and price from the memory and uses this to print out the receipt.
    b) The *proceeds* must be *posted*.
    c) The sold goods are *deducted from the inventory*. If the business falls below the reorder level (see section 5.1.3), then a *repeat order* is due with the supplier or in the central warehouse of a store chain.
    d) Different management information is being prepared for the management of the business.
    e) The sales data permit the creation of customer profiles that may give helpful information about the favorable placement of items or advertising measures.

2. Extra-enterprise
    a) Suppliers with whom corresponding agreements exist receive *information about the sale* of the delivered by the supplier. This information, in turn, may serve for supplier's *observation of the market*, e.g., observing the success of products which are advertised on the local radio at the present time.
    b) Similar information may be passed along to *market research firms*, but the difference is the provided data not only pertain to items from just *one* supplier.

c) In case a POS banking system exists *the amount for payment* is *deducted* from the bank account of the customer (see also section 5.2.8.5).

---

**The following are examples of the North American bar code UPC (Universal Product Code):**

UPC-A                                                            UPC-E

The **UPC A** code is the standard version of the UPC code and has 12 digits. It is also called UPC 12 and is very similar to the EAN code. The **UPC E** code is a short version with 8 digits, always starting with a zero. The UPC code is a numeric code which is able to display digits from **0-9**. Each character consists of two lines and two spaces.

The structure of the UPC A code is as follows:

- The **first digit** of the UPC A code says what the code contains:
  0 - normal UPC Code
  1 - reserved
  2 - articles where the price varies by the weight: for example meat. The code is produced in the store and attached to the article.
  3 - National Drug Code (NDC) and National Health Related Items Code (HRI).
  4 - UPC Code which can be used without format limits
  5 - coupon
  6 - normal UPC Code
  7 - normal UPC Code
  8 - reserved
  9 - reserved

- The **second to sixth** digits show the producer of the article (UPC ID number). This number is issued by the *Uniform Code Council (UUC), 7051 Corporate Way - Suite 201, Dayton, OH 45359-4292, USA.*

- The **seventh to eleventh** digits show the individual article number issued by the producer.

- The **last** digit is the check digit. This one is calculated by ActiveBarcode automatically.

**The following is an example of the European Bar Code, the European Article Number (EAN):**

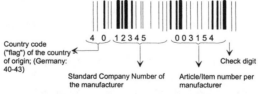

Country code ("flag") of the country of origin; (Germany: 40-43)

Standard Company Number of the manufacturer

Article/Item number per manufacturer

Check digit

---

*Fig. 5.2.8.2/1    Examples of UPC and EAN Bar Codes*

An important element of modern payment systems are cards whose payment function may be designed differently:

■ *Prepaid cards* ("pre-pay") have to be purchased for a particular amount or will need to be pre-loaded prior to their use by the customer with a sufficient amount of money (e.g., telephone cards or chipcards like GeldKarte, see section 5.2.8.3).

- *Debit cards* ("pay now") are tied closely to the customer's bank account. Payments made with debit cards are deducted from the customer's bank account instantly or a few days later (e.g., electronic cash, see section 5.2.8.4).

- *Credit cards* ("pay later") usually offer the customer a certain payment goal, since the card charges are typically collected monthly and will have to be paid by the customer to the credit card-issuing institution only then. Often it is possible to negotiate the mode of payment, e.g., credit limit and payment periods.

Modern chipcards (Smartcards, see section 2.1.2) like GeldKarte have their own processor and are thus capable to be used for additional applications aside from their original payment function. For example, it is possible to use them as student identification cards or to store and implement encryption functions on the chipcards for digital signatures. Beyond that application-specific chipcards are used, e.g., the insurance card with insurance companies and health plans or the SIM (Subscriber Identity Module) card in cellular phone applications.

Secure payment systems on the Internet will enable the buying and payment via data networks. One operates with, e.g., "signed coins" (electronic chits) that are stored on one's hard drive or chipcard. In addition, there are various schemes and standards that make a secure payment transaction through the encryption of credit card numbers and money amounts possible.

It should be mentioned that the creation of electronic money in theory and practice must meet three criteria, just like the so-called real money, i.e., dollars and Euros.

1. Money has to be an accepted medium of exchange. As a society, nation or region, we must agree that the dollar or Euro is an accepted medium of exchange. If this does not occur, as it sometimes happens with currencies from countries which the world financial community has no confidence in, this currency is unlikely to be traded or accepted as a medium of exchange.

2. Money has to have symbolic value. Cash as we know it does in and by itself not really have the value it represents, but we all agree that the printed value on the paper note is real, true and correct.

3. Money must be a medium to store value over time. If one places $100 into a bank account and withdraws it one year later, one usually would expect that the value has at least remained the same; maybe one gained some interest as well. When a currency is rapidly increasing in its de-

nominations, but not in its value (like in extreme inflationary times), this currency would be an undesirable place to store one's value.

The very same tests can and must be applied to electronic money. We may have to use trusted third parties, e.g., the issuer (maybe an Internet bank) of electronic money, i.e. the electronic chits in $10, $20, $50, etc. denominations. This third party would charge the customer's credit card for each chit issued/purchased and at the same time the issuer guarantees the authenticity and value of the chit. When submitting a chit for payment, before shipping the goods purchased the merchant would run these payment chits by the trusted third party (this happens on-the-fly, unnoticed and in the background of the original transaction). The trusted third party would vouch for their authenticity and value, thus guaranteeing payment to the merchant.

Worldwide there are many card payment schemes, several of which are based on the Internet and have been created for this specific use. Numerous payment schemes have been invented (Mondex card, PayPal used by eBay™, eCash, CyberCash, Digicash, Verifone, etc.) but at least in the North American context there is not one electronic card scheme (on or off the Internet) that has developed as a major player.

### 5.2.8.3 Payment with a Pre-paid Card

One of several rechargeable *electronic purses* is the so-called *GeldKarte* that has enjoyed considerable success in the European context. In principle all chipcards and their underlying transaction architecture work quite similarly and we have chosen GeldKarte to illustrate this. GeldKarte, meaning literally 'money card' in German, is a chipcard issued by banks and mutual saving-and-thrift banks. For example, all Eurocheque cards issued by mutual saving-and-thrift banks (Sparkassen) have a chip since 1997.

The payment process may be broken down into the following payment phases (cf., fig. 5.2.8.3/1):

1. The merchant's bank issues a *Merchant Card* that is needed together with an expanded electronic cash terminal in order to participate in the Geld-Karte process. The Merchant Card is needed for authentication of the merchant vis-à-vis the terminal and also for the banking transaction.

2. The customer's bank issues the GeldKarte. It may be recharged at automatic teller machines for up to €200 and can be read by these machines, as well as by the terminals of participating merchants. The customer identifies him- or herself using a *Personal Identification Number* (PIN) that was sent by the bank. The card stores the last 15 funds-spent entries, a feature assisting the customer. The customer's bank opens an exchange settlement account in which the stored value of the original card, the recharged credit amounts, as well as the various debit entries are listed.

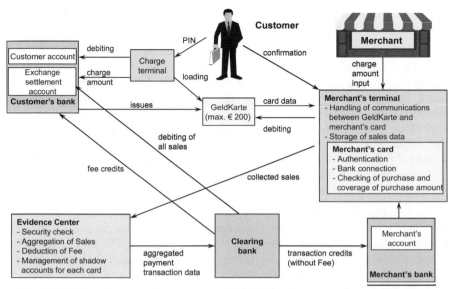

*Fig. 5.2.8.3/1    The underlying system of GeldKarte*

3.  When paying with the GeldKarte, the card is entered into the merchant's terminal. The merchant enters the amount of the actual purchase and the customer confirms this. Several checks occur on-the-fly and in the background of the transaction on the merchant's side, e.g., checking whether or not the purchase amount is covered by the customer's bank account. Finally, the amount is deducted from the customer's GeldKarte chip and is stored within the merchant's terminal device.

4.  The collected sales of the merchant will be transmitted electronically (e.g., using an ISDN line) to the so-called Evidence Center. The center carries out various security checks (e.g., authenticity of the individual transactions and recognizing multiple submissions), aggregates the sales data and transfers the payment transaction data to an associated clearing bank. Aside from these activities, the center manages a shadow account where every card transaction is recorded. With flawed submissions the merchant receives a notification via his/her bank and is being asked to resubmit the data another time.

5.  The clearing bank transmits the amount excluding the associated fees (presently 0.3% of the amount to be paid) to the merchant's account of the merchant's bank, debits the amount from the exchange settlement account of the customer's bank and transfers the fee to the customer's bank.

An advantage of a pre-paid card vs. the use of a debit card during so-called electronic cash transactions (see section 5.2.8.4) is the express warranty of complete payment security, even without an online authorization. In addition, no network operator or other third-party vendor needs to be utilized and the fees to be paid by the merchant are lower.

## 5.2.8.4 Payment with a Debit Card

Debit cards are being used in many parts of the world today. In North America the customer typically will be issued automatically a debit card when opening a new bank account. North American banks are very interested in the use of debit cards as they wean reluctant customers away from the old habit of still writing paper checks, even for relatively small purchase amounts, e. g, paying for a purchase in a grocery store. The feature of using paper checks as part of a customer's bank account is a solidly money-losing and very time- and labor-intensive proposition for banks. Virtually all North American banks are trying hard to get rid of paper checks, but the customers' habit and expectations are tough and slow to break. In principle all debit cards work the same way and similar information system architectures are used worldwide. In the model below we describe the operation of the Eurocheque debit card.

A typical debit card is the Eurocheque card that is used for electronic cash transactions in the European context. Figure 5.2.8.4/1 demonstrates the handling of payments with electronic cash.

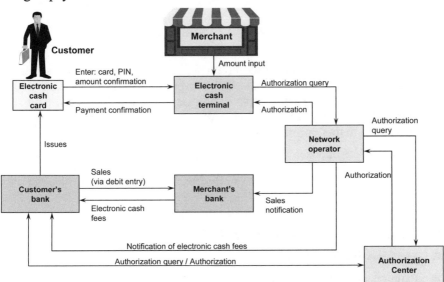

*Fig. 5.2.8.4/1   Electronic Cash*

The payment process can be classified into the following phases:

1. The customer banks issue electronic cash cards needed for the use of electronic cash, partially still equipped with magnetic strip and PIN.

2. At the merchant's business the customer identifies him-/herself by inserting the card into the POS terminal and by entering the PIN. Then the customer confirms the due amount to be paid which is determined by the POS system. An authorization query is then sent via the network to the *authorization center*. The center examines the authenticity of the customer, as well as that the account has sufficient funds and that no blocking of the account has occurred. The authorization occurs in the same way, but in reverse order. The customer receives at the terminal the message that payment will occur and a paper sales slip will be issued for this electronic transaction.

3. In terms of handling the payments, either the merchant submits the sales data via *data carrier exchange* or electronically to the network operator or the latter takes on the compilation of the transacted sales. In turn, the aggregated sales data are messaged to the merchant's bank and the electronic cash fees are transmitted to the buyer's bank.

4. Payment occurs by collecting the sales amount using the debit advice procedure via the merchant's bank from the buyer's bank and the collection of the electronic cash fees, but in the reverse order as described previously. The buyer's bank received the fees, as it guarantees the payment for their customers.

### 5.2.8.5    Cashless Transaction Systems of Banks

Banks are important enterprises within the financial services sector, offering products like money investments, credit and financial transaction services. Accordingly, they offer complete service processes with the corresponding phases. The product 'monetary transaction' is being utilized in the payment phase of nearly all services in various industries by private customers, as well as companies. In this, information systems may support two areas:

- The interface between the customer and the bank
- The actual processing consisting of handling, transmitting and clearing activities

    Basic principles of progressive monetary transactions are:

- One-time capture of the monetary transaction
- Machine-based processing
- Paperless, electronic transaction
- Worldwide networked integration
- Automatic printing and shipment of paper receipts

    Figure 5.2.8.5/1 depicts the scope of functions of monetary transaction systems in combination with a few examples.

For the efficient processing of the increasing number of receipt slips, receipt recognition and sorting machines have been used early. In the European context the receipt slip-based orders are declining in the meantime; the automatic entry and transmittal are standard. This is different in North America, where banks are working very hard on this issue. Although so-called electronic banking has made some in-roads in North America, its success is still in its infancy. Moreover, the systems offered by banks in North America are no electronic bill payments in the generic sense, because electronic bill payment is only possible with those merchants and firms that the bank signed up for this specific purpose. It is not a generic bank funds transfer system (true electronic banking) as has been in existence for many decades in Europe.

The European setting uses money transfer forms to be completed by the bank account's owner in order to transfer funds from the account holder's account to account holders in Europe. Such forms are read by OCR receipt readers, hand writing reader devices or scanners (see section 2.1.4) and, if needed, may be manually entered. Within the computer center of the sender's bank occurs, e.g., with funds transfers the debit entry (as part of account data management); data sets for the electronic monetary transactions are generated. After transmitting the funds transfer notice via the bank communication network the credit entry occurs within the computer center of the receiving bank, here, just the same, as part of account data management. Even today, the customer may print out his/her account statements at particular terminals at the bank. It should also be noted that the entire process described above may be done electronically using the bank's web-based personal account management features, i.e., the transfer form may be completed and activated from the bank's website by the account owner.

With a *data carrier exchange* the application systems of a company's customers no longer create printed receipts such as with payroll, invoicing or accounting programs. The output data are written by a monetary transaction program as a file onto a data medium such as diskette or magnetic tape.

The contents of the arriving data carriers are read by the bank, are checked and transmitted to the computer center, are compiled there and then transformed into financial transactions. This is referred to as magnetic tape or diskette clearing.

When the orders for financial transactions are already generated electronically at the customer interface and are then fed into the electronic monetary transaction systems of the banks, one refers to this as the Electronic Funds Transfer System (EFTS). Various systems have been developed that are orientated according to customer needs.

Access to EFTS may occur via self-service terminals, via the Internet or online services such as home banking, but also via bank terminals. In retail stores we find electronic cash systems at the POS (see section 5.2.8.4) at the customer's disposal.

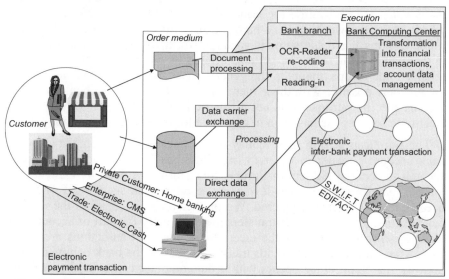

*Fig. 5.2.8.5/1    Computer-supported Payment Transactions*

In the following all payments are processed via the closed monetary transaction network of the banks. The structure of the bank routing system in the United States is described in Figure 5.2.8.5/2. The *basic routing number*, devised by the *American Bankers Association* (ABA) in 1910, has served to identify the specific financial institution responsible for the payment of a negotiable instrument. Originally designed to identify only check processing endpoints, the routing number system has evolved to designate participants in automated clearing centers, electronic funds transfer, and online banking. The ABA Routing Number (a.k.a. ABA Number; Routing Transit Number) has changed over the years to accommodate things like the Federal Reserve System, the advent of MICR, and the implementation of the Expedited Funds Availability Act (EFAA). A routing number will only be issued to a Federal or State chartered financial institution which is eligible to maintain an account at a Federal Reserve Bank.

For the processing of monetary transactions between and among banks the SWIFT (Society for Worldwide Interbank Financial Telecommunication) system or EDIFACT-Messaging is being used (see section 5.1.2.3).

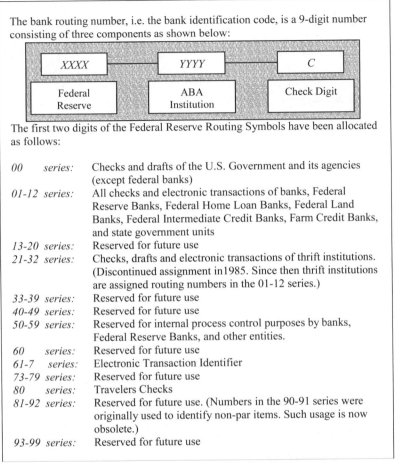

The bank routing number, i.e. the bank identification code, is a 9-digit number consisting of three components as shown below:

The first two digits of the Federal Reserve Routing Symbols have been allocated as follows:

| | | |
|---|---|---|
| *00 series:* | Checks and drafts of the U.S. Government and its agencies (except federal banks) | |
| *01-12 series:* | All checks and electronic transactions of banks, Federal Reserve Banks, Federal Home Loan Banks, Federal Land Banks, Federal Intermediate Credit Banks, Farm Credit Banks, and state government units | |
| *13-20 series:* | Reserved for future use | |
| *21-32 series:* | Checks, drafts and electronic transactions of thrift institutions. (Discontinued assignment in1985. Since then thrift institutions are assigned routing numbers in the 01-12 series.) | |
| *33-39 series:* | Reserved for future use | |
| *40-49 series:* | Reserved for future use | |
| *50-59 series:* | Reserved for internal process control purposes by banks, Federal Reserve Banks, and other entities. | |
| *60 series:* | Reserved for future use | |
| *61-7 series:* | Electronic Transaction Identifier | |
| *73-79 series:* | Reserved for future use | |
| *80 series:* | Travelers Checks | |
| *81-92 series:* | Reserved for future use. (Numbers in the 90-91 series were originally used to identify non-par items. Such usage is now obsolete.) | |
| *93-99 series:* | Reserved for future use | |

*Fig. 5.2.8.5/2    Structure of the Bank Routing Number System*

## 5.3   Electronic Commerce

### 5.3.1   Overview

Product sales, as well as aspects of customer relationship management may be supported effectively using the new electronic media, especially the Internet or the WWW (see section 5.2.3.2). The corresponding interactions and business transactions between the company and the customer are referred to as electronic commerce or e-commerce.

Not all products are equally well suited for the distribution via electronic communication media. After the initiation and agreement phase electronic support of the service or performance delivery during the execution phase depends on the degree to which the delivered service is digitizable. Electronic processing is not possible for physical goods and certain services that require a personal, i.e. often face-to-face, interaction between employees of the company and the customer or the service object (e.g., an automobile in need for repair). Digitizable services are suited for electronic processing, if products are of a nature such that information is the essential benefit factor for the customer or the actual service and performance result itself.

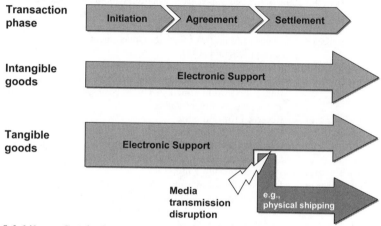

Fig. 5.3.1/1    *Goods that may or may not be digitized*

This is the case with many services such as consulting services. The transaction may then be completely carried out via electronic networks (cf., fig. 5.3.1/1).

| Phase | Sub-Task | Support |
|---|---|---|
| Initiation | Public Relations | Website |
| | Advertising | Newsletter |
| | Product catalogue | Website with database connection |
| | Consultation | Video conferencing, Frequently Asked Questions (FAQ) with WWW |
| Agreement | Product configuration | Online configuration system |
| | Ordering | WWW forms, Electronic Data Interchange (EDI) |
| Execution | Delivery | Download of digital products |
| | Payment | Digital money, Smartcards |
| | Customer Support | Hand books, Frequently Asked Questions (FAQ) with WWW |

Fig. 5.3.1/2    *Examples of Electronic Support in Transaction Phases*

One should note though that with non-digitizable products as well the first two transaction phases may be largely supported electronically. Exclusive electronic processing, however, is rare. Figure 5.3.1/2 provides an overview about the technical possibilities.

## 5.3.2 Flow of an E-Commerce Business Transaction

The basis of an e-commerce solution for electronic product sales is typically the use of a shopping system on a web server. Here private and corporate customers may retrieve information, may place orders and they may pay for what they ordered.

During the initiation phase the customer has the possibility to look at the offered goods in the form of an electronic product catalogue and to place the chosen products into an electronic shopping cart. Here the goods are described and explained using text and photos. Three-dimensional depictions illustrate more complex products such as cars and furniture using Virtual Reality Modeling Language (VRML). In this way, i.e. through a three-dimensional perspective, the customer has the possibility to make his/her selection decision a little easier.

Often a search function is being offered through which the user may search for certain products, brands or using other key words within an entire line of goods. In addition hints for similar or complementary goods may be offered in the description of individual goods. For example, during the purchase of a pair of pants the system may offer automatically a matching belt. Progressive support of the customer may occur through software agents of the seller (see section 4.3.2.3) that search for suitable products and recommend these based on the customer's broad specification of wishes and desires. The mail order catalogue and web-based company Lands' End™ allows the buyer to try on different clothes and find the, e.g., best color choice by building a model, i.e. the customer's model, by specifying five basic features of one's physique. Thus he may see if, e.g., the blue skirt matches the beige color of the sweater. Moreover, he can invite a friend to shop together 'interactively', compare one's selections and ask the friend's opinions about his choices and preferences. The friend may sign in for this joint shopping session from anywhere where a WWW access is available. Further it is possible to store self-configured models on Lands' End's website so that they are retrievable at one's next visit. It is possible to provide returning customers offers on the homepage that are specifically tailored to this customer, based on past activities, items purchased, interests and preferences. The underlying principle in offering such a service is based on the comparison of the focal customer when compared to the behavior of other customers with a similar buying profile (*collaborative filtering*).

Customers may find support by using lists of frequently asked questions (FAQ), as well as the possibility to get advice from employees of the seller.

Aside from email contact special "call-me-buttons" will trigger a phone call by an employee. Customers may be recognized by *Cookies* (very small data files that are stored by the web server of the seller onto the customer's computer and may be recalled during a repeat visit by the customer to the same website) or right at the point in time when they log in repeatedly at the same website.

Before completing the purchase the customer gets the opportunity to check the selected goods in the shopping cart one more time. Features enabling barters or negotiations about prices are rare. The shopping system typically merely calculates the purchase price for the entire order that has to be accepted by the customer. Depending on the seller shipping and handling fees, as well as sales tax may be added, depending on if the seller has a physical "presence" or "nexus" (meaning store, office or subsidiary) in the state from which the purchase is made.

Payment just as the ordering portion may occur electronically. So far it is common to use conventional means to buy. In the North American context that is typically the credit card. In Europe, credit cards have gained widespread usage, but it is still common to pay by making funds transfers (see section 5.2.8.5). It is also not unusual to use the cash-on-delivery (COD) method of payment, i.e. the postal service collects the money when delivering the shipment at the customer's home. So far as the sold products cannot be digitized, the physical shipment of the goods constitutes the last step of the process. The order will be placed automatically into the material management system and will be processed. Customers may follow today with many companies the production and shipping process through the popular *tracking and tracing* feature. Moreover, the automatic sending of status information via electronic mail permits the customer to observe the correct processing of the order. From the seller's perspective, the feature is attractive because the customer does all the work (almost a form of free outsourcing) and it frees up considerable resources on the seller's side that otherwise would have to be activated (e.g., increased staff to answer free 800 number calling services). Federal Express reported in October 2002 that it saves $25 million per month with this tracking feature, as opposed to the 800 number calling alternative.

# 5.4 Integration of Enterprises through Supply Chain Management

Coming to the close of this chapter entitled, *"Integrated Application Systems"*, we would like to offer the reader an important example of interorganizational integration in which production and service businesses (trade, logistics-service providers) are participating: With *Supply Chain Management* (SCM) one attempts that all participating parties in a *supply chain* or a *supply network* (cf., fig. 5.4/1), respectively, gain an advantage (a win-win situation).

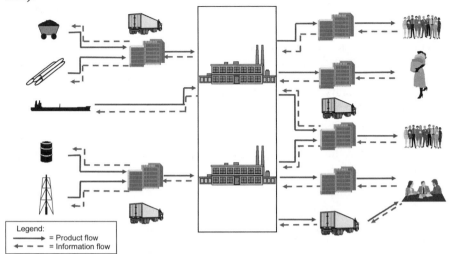

Legend:
⟶ = Product flow
←  −  − = Information flow

*Fig. 5.4/1     Product and Information Flows in the Logistics Chain (Modification based on [Mertens 01, p. 295])*

*A PRACTICAL EXAMPLE*

*A retailer plans a sales promotion for an item. Based on the stored data from an earlier promotion the system estimates that sales will increase by 50% above the normal level. This forecast is being coordinated with the supplier. The supplier retrieves the company's historical time series data and adds external data that the company had bought from a market research firm. Then the sales system of the supplier predicts that due to the promotion also neighboring stores will experience an increased demand. The supplier recommends that other partners should modify their sales predictions and the subsequent plans (based on these predictions) for production and additional warehouse storage.*

An important element of the SCM concept is that sales forecasts of individual partners are being substituted by exact information about sales, warehouse inventories and order dates, as well as order quantities of the businesses that "lie on the road to the end-consumer". In the extreme case the computers of all partners, i.e. the producers, their suppliers, two-tier-suppliers, the warehouses, wholesalers, shipping agents, etc., get the data transmitted that were registered by the retailer's POS system (see section 5.2.8.2). In this way they find out what has been sold "at the end-consumer front". For this purpose the participants also permit that their computers gain access to the corresponding and appropriate databases. *Cooperative Planning, Forecasting and Replenishment* (CPFR) could be the starting point for SCM. This means the partners provide each other with their demand forecasts. These predictions are achieved using computer-supported procedures such as exponential smoothing (see section 5.1.3.1). An information system calculates from this a joint forecast, in the simplest case by using the mean. From such demand estimates schedules are derived by using parts lists explosions (see section 5.1.5.3), lot formation (see section 5.1.3), and similar production, planning and control elements.

Through the use of complicated rules the system examines alternatives how a demand may be met in a timely fashion (available-to-promise technique (ATP)). Such alternatives are assembly by locally available and ready main building groups, substitution of a non-available product by another one, procurement from a distant location, e.g., from a distribution center in Singapore. In the latter case sophisticated SCM software will assure that a cargo airplane with available loading capacity will fly from Singapore in a timely fashion. ATP may be viewed as a continuation of the availability check (see section 5.1.5.6) in an interorganizational setting.

Possibly the supplier schedules the warehouse of each respective buyer (*Vendor Managed Inventory, VMI,* cf., also [Knolmeyer et al. 01]).

We would like to thank the following individuals for their information on the current status of application systems in their companies: Mr. A. Kerl (Dynamit Nobel Kunststoff GmbH, Weissenburg, Germany), Mr. H. Pirner (INA Waelzlager Schaeffler oHG, Herzogenaurach, Germany), Dr. M. Schmidt (Carl Zeiss, Oberkochen, Germany) and Mr. T. Wedel (IBM Deutschland GmbH).

## 5.5   Literature for Chapter 5

Banks 01 — Banks, E., e-Finance, The Electronic Revolution, John Wiley & Sons, New York 2001.

Bodendorf 99 — Bodendorf, F., Wirtschaftsinformatik im Dienstleistungsbereich, Berlin 1999.

Lovelock 00 — Lovelock, C. H., Services Marketing, Prentice Hall 2000.

Feldmann et al. 98 — Feldmann, H.-W., Droth, D. and Nachtrab, R., Personal- und Arbeitszeitplanung mit SP-Expert, WIRTSCHAFTS-INFORMATIK 40 (1998) 2, pp. 142-149.

Haberl 96 — Haberl, D., Hochleistungs-Kommissionierung im Kosmetikunternehmen, VDI Berichte (1996) 1263, pp. 93-138.

Knolmeyer et al. 01 — Knolmeyer, G., Mertens, P. and Zeier, A., Supply Chain Management based on SAP systems – Order Management in Manufacturing Companies, Berlin 2001.

Krill 96 — Krill, O., EDI – eine Voraussetzung für eine sequenzgerechte Belieferung der Automobilindustrie, Industrie Management 12 (1996) 6, pp. 47-50.

Mertens 94 — Mertens, P. (Eds.), Prognoserechnung, 5th edition, Würzburg-Wien 1994.

| Mertens 00 | Mertens, P., Integrierte Informationsverarbeitung 1, Administrations- und Dispositionssysteme in der Industrie, 12th edition, Wiesbaden, Germany 2000. |
|---|---|
| Mertens/Griese 02 | Mertens, P. and Griese, J., Integrierte Informationsverarbeitung 2, Planungs- und Kontrollsysteme in der Industrie, 9th edition, Wiesbaden, Germany 2002. |
| Mertens et al. 94 | Mertens, P. and Morschheuser, S., Stufen der Integration von Daten- und Dokumentenverarbeitung - dargestellt am Beispiel eines Maschinenbauunternehmens, WIRTSCHAFTSINFORMATIK 36 (1994) 5, pp. 444-454. |
| O'Mahony 01 | O'Mahony, D. et al., Electronic Payment Systems for E-Commerce, Artech House, 2nd edition, 2001. |
| Österle et al. 00 | Österle, H., Fleisch, E., Alt, R. (Editor), Electronic Business Networking: Shaping Enterprise Relationships on the Internet, Berlin 2000. |
| Rust et al. 02 | Rust, R. T. et al., E-Service: New Directions in Theory and Practice, M.E.Sharpe 2002. |
| Scheer 90 | Scheer, A.-W., Computer Integrated Manufacturing – der computergesteuerte Industriebetrieb, 4th edition, Berlin 1990. |
| Zerdick 00 | Zerdick, A. et al., E-conomics: Strategies for the Digital Marketplace, Springer Verlag, Heidelberg 2000. |

# 6 Planning, Implementation and Deployment of Application Systems

Application systems (AS) are to support the user effectively in handling business processes and management functions. For many areas we can find suitable software packages meeting the demands of the job to be done. If, on the other hand, the needs of a firm are very specific we then have to develop custom software solutions or the *software package* has to be modified or expanded, respectively.

Future users from the various *functional departments* will have to define how and in which form their work needs to be supported. The chosen *software package* or the application system to be developed has to meet these demands.

Aside from the *subject matter conceptualization*, experts involved with *systems development* and possessing special information systems knowledge have to conceptualize and implement the application system from an *information technology perspective*.

Within the framework of systems development one often also encounters the rather technical term *software engineering*. Under this term we understand the knowledge and application of principles, methods and tools for the creation and maintenance of software, as well as the related management tasks.

## 6.1 Fundamental Decision: Software Packages versus Individual Software

Generally, the decision whether or not to use *software packages* vs. *individual software* may be made after analyzing the application area and the information systems environment. Some firms may have rules that for certain tasks only packaged software may be used as a basic principle.

A prerequisite for the deployment of a *software package* is that the demands of the firm have to largely coincide with the performance characteristics of the software package available on the market. The firm must also be able to abstain from partial functions or to implement those itself separately. Moreover, the firm must realize that possibly some form of *organization change* may be necessary in some areas due to the software package implementation.

Following arguments for and against software package usage are presented. Advantages of software packages are at the same time disadvantages of individual solutions and vice-versa.

- Advantages:
    - Usually the costs for acquiring and adapting software packages are lower than the costs for the building of a custom solution.
    - Since standard packages are available immediately the duration of the implementation is usually much shorter than with custom software that will need to be developed in the first place.
    - Software packages are often mature software such that fewer mistakes than with custom software are likely to occur.
    - With software packages it may be possible to acquire management and organizational know-how that presently may not be available within the firm. An example is a new production, planning and control system enabling improved capacity planning for manufacturing and will make it possible for sales to inform customers more quickly about planned delivery dates.
    - Integrated software packages permit the simple connection of differing management tasks, as well as with suppliers and customers.
    - Widely distributed software packages simplify interorganizational integration if generally accepted standards are being used.
    - Often training that is offered by a software vendor for users is more professional than training offered by an in-house department of a firm.
    - A firm's own information systems resources are conserved in a way that they may be used for particularly important tasks (see section 7.2.1).

- Disadvantages:
    - Often there are discrepancies between functional and organizational management demands and program structure.
    - Unit- or individual-specific software modifications may bring about high expenses, especially when one considers that these changes will have to be repeated anew with a version change of the software package.
    - Hardware is strained additionally, as the software is not customized to the specific computing environment of the firm.
    - Little internal information systems know-how is being developed in the focal firm.
    - The firm may end up possibly in an unwanted dependency relationship with the software supplier.

■   The adaptation of the software package to the realities of the user's firm occurs usually via parameter settings with effects on planning and scheduling processes that are often difficult to foresee.

In order that the discrepancies between management demands and the functional complexity of software packages will not become too great, software designers offer solutions through which firms may choose functions or modules as various alternatives. In this case programs are used for combining individual modules to the final application software (see section 6.2.3.2).

As a newer development it is discussed to use application software neither as full custom software nor as a software package. The underlying idea is to build application software from software building blocks within which the needed technical requirements may be found. Therefore, modules are used that fulfill special tasks and that become reusable on the basis of standardized interfaces. One anticipates that such a *component architecture* or *component ware* promises a better adaptation to the business idiosyncrasies while at the same time utilizing advantages of the deployment of packaged software.

# 6.2   Structuring of Projects

The new and further development of application software, as well as the implementation of software packages occurs within a projects framework. The contents vary, with the exception of raising technical demands and certain tests. In contrast, widespread acceptance is found in the methods for project management (see section 6.3).

For the introduction of software packages, as well as for software development in general the development life cycle methodology is being used that partitions the project into partial steps. When using the development life cycle methodology for software development, one assumes that a program will be structured and refined, based on an exact problem delineation and task specification, in a step-by-step fashion until the individual commands get coded. Analogous to the field of technical engineering, this way of proceeding results in run-capable software to be generated not until the latter phases of the development process. The alternative is the faster development of an initially run-capable prototype for which an exact specification does not play an outstanding role. Not until feedback is received from some users one will decide whether at all or in which form the prototype is to be utilized.

## 6.2.1 System Development Life Cycle for Custom Software

In the development life cycle methodology the development process for an application system is dissected into subsequent steps. Individual partial steps come to fruition each time by having deliverable results that also constitute the input for the very next phase. One refers to this way of proceeding as the waterfall model. Based on the partial results a quality control test may be conducted. If one realizes in a subsequent step that tasks based on decisions in previous phases were not solved satisfactorily, one will have to back-track to the phase where the decision was made and that led to this current problem. Getting rid of errors in this fashion may become rather costly.

In the literature one finds many references to numerous versions of the development life cycle methodology that differ mainly by their way of labeling the partial steps and their delineation of the content of the steps. As an example a process of six steps is outlined [Martin 02, p. 15 et sqq.]:

1.  Planning step
2.  Definition step
3.  Design step
4.  Implementation step
5.  Acceptance and introduction step
6.  Maintenance step

Parallel to these six steps one should undertake the effort of a permanent documentation to capture the results of the individual life cycle phases.

### 6.2.1.1    Planning Step

In the *planning step*, the targeted goals and the rough functional scope of the application system are described based on the project idea and the outlined content. In a first preliminary inspection the technical feasibility and cost effectiveness are analyzed.

The technical feasibility pertains, e.g., on the assessment of the usability of already available hard- and software, whether an already available database may be deployed or if the software to be developed will reach the desired performance level. Here, the programming paradigms (see section 2.2.1.2) for the system components have to be chosen as well. In the cost effectiveness examination one needs to determine the development costs (see section 6.3.2), as well as the efficiency gains (see section 7.1.3.2).

In efforts to standardize the software development process firms often merely deploy a development environment or few varying tools (see section 6.4.2). For new developments one often uses C++ or Java. Especially for programs that run on mainframes COBOL is still widely used. In the development environment for Internet applications, aside from Java there can also be found interpreted script languages such as PERL, TCL and PHP (see section 2.2.1.2) that recently have been enjoying increasing usage.

### 6.2.1.2    Definition Step

In the definition step the demands for the system to be created are defined, primarily from the perspective of each involved department. For this a *cur-*

*rent condition assessment* is conducted of the area in which the application system will be introduced. Frequently, the term *requirements engineering* is used in this context.

One of the central results of the definition phase are requirement specifications that more or less formally specify the demands on the system and the development process. Depending on the complexity of the task the activities can be repeated in multiple cycles in order to compile the requirement specifications. The main difficulty when describing requirements is often that the understanding, i.e. eventually, the shared meaning, between later users and information systems-oriented system developers diverges. The employee from the specialist department within a functional area is used to think within an applications-focused world, whereas employees of the information systems department are rather technically oriented. Therefore, a communicable form of expression has to be found that is acceptable to both sides.

The demands on the new system evolve through an analysis of the current condition and the identified weak spots, as well as the general goals of the firm (see section 7.1.1). The results are to be documented in a *blueprint* within the framework of the requirements document.

In the analysis of the current condition the first step is to determine the area of investigation in which the analysis is to be conducted (Example: In order to improve a warehousing system should only the materials movement be examined or must the inventory be considered as well?). In the second step the current, actual condition within the area of investigation needs to be determined. It is helpful here to examine the broad context and relationships first and then to delve into details (top-down approach; see section 6.4.1.3). The systems analysis may be less extensive when we already know that sizable changes are planned.

Examples of instruments for initiating the systems analysis are:

- Interviews (structured and unstructured) in which the employees are interviewed in the field to be analyzed

- Workshops in which the demands of the potential users of an application system are assessed

- Questionnaires that serve to gain information about pre-specified themes and topics

- Observations of employees carrying out actual tasks

- Documentation analysis (e.g., a study of the forms utilized in a business process) in order to reconstruct processes partially

- Write-ups or protocols in which for example employees record work processes and the tools utilized

The developed demands for the requirements specification of the software product may be differentiated along functional aspects, quality aspects, as well as economical aspects.

Functional aspects describe:

- Scope of the function that has to be met (e.g., a new system for travel route scheduling is to be designed; should it plan only the round-trip travel route or should also the conveyance papers for the vehicles be printed out?)

- The Way in which the application system generates functions (e.g., information for a management support system will first have to be loaded into a PC database such that it can then be evaluated and presented individually there)

- The data structures for the functions

- In- and outputs of systems, as well as their interrelationships

- Design of the user interface

- Expectations about the response times

- Reliability of the system

Economic aspects pertain to later management and maintenance costs. All results are to be documented as accurately as possible in the requirements document in order to describe the demands of the software for its practical deployment.

Parallel to this effort we need to specify the development process. Functional perspectives here are, e.g., the required collaboration between the individual specialist departments with project tasks, activities during system introduction or change, as well as the maintenance after initial operation. Guidelines for quality influence program documentation or software testing efforts. Economic demands pertain to costs, the duration, as well as the actual usage of various development resources.

## 6.2.1.3    Design Step

The requirements document is the basis for the subsequent *design step*. Our goal is to describe the entire application system as a hierarchy of largely mutual independently developed and reusable partial systems (modules). In doing so one initially assumes that, based on recent technical progress, computer equipment and networks are of sufficient high-performance caliber such that a *technology-independent design* gets created. Subsequently, this design is passed along into a *technical design of information systems*.

The specification of the individual modules and their combination to a unified and complete system depends on the form of implementation to be used in subsequent steps.

The essential goal with the technical design of application systems is to carve or peel out the functions, as well as their relationship and the data to be processed. In doing so we need to examine how the relevant data are being generated, used, actualized, deleted and exchanged. Results of the technical design may be, e.g., data and functional models (see sections 6.4.1.2 and 6.4.1.3). These show the data and functions that are required in order to handle the processes of the firm, to prepare decision tools for executives or to generally achieve the firm's goals. If, on the other hand, the technical design is undertaken using the object-oriented approach then these „views" of the management reality are being unified in the object model (see section 6.4.1.4).

The technical design of information systems builds up from the technical specifications and considers the environmental conditions pertaining to hardware, system software or the programming language to be utilized. While the database model, e.g., may be transformed directly into a relational database model (see section 3.1.8.1), the functional model may not be converted directly into a program module. The reason for this is that, e.g., multiply reusable modules should be specified only once or that building blocks need to be designed that do not accomplish technical tasks directly (e.g., data access, error handling). The goal of object-oriented modeling as a variant of data and functions modeling is, however, to take on the technical design largely through complementing the management descriptions (see section 6.4.1.4).

An additional problem of increasing importance may be found in the area of distributed applications, i.e. those running on several computers that communicate among each other via networks. In doing so the trend is shifting from once monolithic approaches ranging from client-server applications (see section 2.4.4) to multi-level architectures (N-tier architecture). With Web-based application systems one uses today three or more layers that take on varying functions (e.g., presentation, applications-logic and data administration layers). These layers are logically and/or physically separated and communicate merely through defined, usually network-based interfaces. For large applications it is therefore necessary to create a concept about the distribution of functional elements on the existing or planned hardware resources, as well as the necessary communication possibilities.

The result of this phase during a conventional development process will yield:

- Overall structure (components) of the application system (e.g., in an order system a component for quantity scheduling and a component for supplier selection) and its distribution (e.g., onto client and server)

- Program modules through which business management functions are implemented (e.g., for the supplier selection there is one module that selects all suppliers, as well as two additional ones that carry out condition and quality comparisons)

- Sequences in which the individual modules are to work within the program

- Logical data structure of the application

- Physical data and database structure

- First test cases

### 6.2.1.4   Implementation Step

The *implementation step* serves to detail the system design up to the level of individual commands and to translate into the chosen programming language. With a *granular concept* the following is determined:

- Data schemata (data structure, data or database descriptions; see sections 6.4.1.2 and 3.1.8) or classes and attributes in the OO case, respectively

- Program flow or functions or information flow in the OO case, respectively

- User interfaces

Subsequently, they need to be coded. One tries to utilize information systems-supported description tools in order to work with so-called *program generators* to create an executable code in the chosen programming language as far as possible without human assistance. In doing so the productivity of the programmer is improved.

For the design of the user platform one has to consider above all the demands of ergonomics. Aside from hardware ergonomics *software ergonomics* is relevant, as it concerns itself with the adaptation of programs to human beings. It analyzes, e.g., which assistance one requires from a text processing system in order to find an unknown command. For dialogue design we should be aware of the standard ISO 9241 in which we can find, e.g., criteria for the task requirements, usability performance and satisfaction, or the clarity for computer screen-based dialogues.

In conjunction with Web applications specialized designer agencies are frequently utilized that focus on the design of the pages and user platforms. In comparison to classical applications here a deliberately greater importance is given to layout-oriented aspects (e.g., typography, graphical elements).

The system test is also a component of the implementation phase. The complete application and individual partial programs building on this are tested carefully. Tests may be characterized using the following elements [Trauboth 96]:

- Objects or tasks (e.g., programs, modules) to be tested

- Development phases during which individual testing may be conducted

- Co-workers (groups, teams) through whom tests may be conducted

- Test types and individual activities that are to be carried out

- Techniques to be utilized during testing

Testing types may be classified as follows:

- Whether one compares the created results based on known test cases only (black box test) or whether one replicates also the program sequence (white box test),

- Whether one carries out a test with specified test objects, a test with coincidental test objects or a weak spots-oriented test with especially error-prone test objects.

Many problems in connection with quality assurance occur when reciprocal dependencies exist and a portion of the quality characteristics is also dependent on the judgment of the users.

## 6.2.1.5    Acceptance and Introduction Step

During the *acceptance and introduction step* one examines whether the program meets the demands of the requirements document. The buyers of the application system may partially prescribe which design methods and procedures have to be used during software development (e.g., testing procedures). This means that, similar to the production of complex goods (e.g., in engineering), not only the finished product, but also the production process documented in the protocol may be subject for review and scrutiny. Then the software is put to work. An important prerequisite for this is user training at an early stage.

## 6.2.1.6    Maintenance Step

Finally, during the *maintenance step,* necessary program changes and adaptations are carried out. One gets rid of errors that remained unrecognized in spite of system tests or that only surfaced after a longer usage of the programs. User wishes and preferences may change as well over time making adaptation measure necessary. Moreover, legal changes may occur, e.g., changes in the tax law that need to be considered in payroll accounting. Maintenance issues may arise when programs require change due to developments in the system environment (e.g., new computers, system software or network components). Research has demonstrated that the maintenance phase, lasting many years until the software is replaced, commands more than 50 percent of the total expenditures of all software lifecycle phases (from the application idea to the disposal of the software).

Following we address those points that are important for the development process, but that cannot be assigned to a specific development step per se.

### 6.2.1.7    Life-cycle-spanning Characteristics

Demands on quality during software development are to contribute that the development process, as well as the software product will exhibit certain characteristics. Measures along those lines may be taken during the design by the functional area specialists. When considering the productivity of the application system, characteristics such as user friendliness, and a commensurate range of functions are of importance, as well as suitable problem solving procedures, the maintainability of the software or the essential and minimal hardware equipment. By-and-large, these are subjective factors.

ISO 9000 offers specific guidelines for software development [Mellis/Stelzer 99]. It contains, e.g., demands for the organizational classification of quality assurance systems, development phase-dependent quality goals and measures for goal attainment, as well as phase-spanning quality-related activities specified within a quality assurance plan. Firms that organize the development process in a way that it conforms with established norms may get their quality assurance system certified by independent evaluators. In this way they can demonstrate to their customers that they meet the specified quality guidelines.

Although these phase concepts have proven useful in software projects over time, we may identify some disadvantages. For example, we assert that at the beginning of the cycle a *complete* and *contradiction-free system specification* is successfully developed. Errors made during that phase may not surface and be identified until later phases. This may delay the development project substantially. Also, the project duration until the first useful software emerges is often too long. Frequently, communication between the information systems department and the functional departments is not satisfactory. This may occur if the future users participate in the development process only during definition phases. During the introduction of the product one sometimes realizes that not all of the users' wishes and preferences were delivered or that user demands have changed in the meantime. This is why one attempts to utilize life-cycle-spanning partial projects in order to recognize mistakes early on. User interfaces, e.g., are often defined as a component of the requirements document. Besides, variants of the phase concept may be used, although these will not be addressed here.

## 6.2.2 Prototyping for Individual Software

During prototyping, two goals are being pursued: On the one hand, one attempts to create a running version of the application software or parts of it early on, without having to conduct a comprehensive problem analysis and without developing a fully-fledged system concept. On the other hand, one attempts to involve the future users to a greater extent during the entire development process.

The prototype is developed in intensive *cooperation between the system developers and employees of the functional departments*. Its first version may only feature selected system functions which are simulated from a user's point of view. Based on this first approach the total system is developed step-by-step, including a continuous, tight working relationship between functional department and software developers. This process is also called *evolutionary software development*. Through intensive participation of the functional departments in these efforts one hopes to achieve a higher rate of acceptance during the following system implementation phase, as well as a reduction of changes and alterations needed to meet additional user demands. A disadvantage of this process is that prototypes do not meet the structural demands of software engineering.

It is possible to combine conventional life cycle concepts and prototyping approaches. Prototyping may begin in the definition phase or during the design phase. Using a prototype emanating from laboratory conditions one would like to gain experiences quickly about the future application software. These experiences may be used in the actual system development as part of the requirements document.

## 6.2.3 Phase Model for Software Packages

If the planning process demonstrated that an integrated software package should be deployed, the firm may either decide to specify a date on which it switches to the new solution completely, or put the software to work successively (e.g., in modules). During a comprehensive exchange of software considerable human resources are required such that all necessary changes may occur simultaneously in all areas. In such projects, the risk of failure is especially high due to the project's complexity and magnitude. However, if the software package is introduced step-by-step, additional expenditures occur compared to the full-blown change. This may, e.g., be due to interfaces to the old systems that have to be created. Since the changes only affect certain areas, the overall risk of these projects is smaller.

Projects for the introduction of software packages usually last several months, whereby the costs for the introduction, especially for human resources, will clearly exceed the costs of the actual software. Usually external consultants who have specialized on the installation of a specific software package are employed in such introduction projects. Aside of the consultants, co-workers from the information systems department—who will be running the application software—as well as *key users* from the functional departments who are responsible for the conceptual specification all work closely together.

Projects for the introduction of software packages take a course that is comparable to the life cycle of individual software development. It may be illustrated as follows:

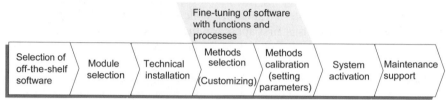

*Fig. 6.2.3/1      Software Implementation Phases*

An entire line of business consultants which has specialized on the introduction of the widely used business management software from SAP has been established. Many of them have specialized on particular modules of the system and use different process methodologies.

In contrast to large companies, mid-sized enterprises cannot afford external consultants for individualized installations and customization. Neither can they afford to employ specialists as regular employees. Consequently, the software manufacturers attempt to pre-define software setups having the needs of such organizations in mind. The software, which has largely been standardized to suit mid-sized firms' needs, may then be implemented without the need for further customization.

### 6.2.3.1     Selection and Installation

The requirements document and the requirements catalogue are utilized for the selection of software packages. Vendors will then have to demonstrate to what degree they can fulfill these specifications. Vendor capability may also be checked by using test installations or by contacting reference customers.

Some guidance may be provided by comparing the relevant business processes within the company with the process models documented by the producer of the software package.

The selection of relevant software modules (e.g., materials management, sales and distribution, financial management, etc.) may also be determined by the business processes and functions needing support. Next, the software that has been selected as described above has to be installed.

## 6.2.3.2    Adaptation and Initialization

As soon as the software is installed, the modules need to be adapted to the characteristics of the functions and processes. In order to do this, suitable methods (possibly available from an inventory of methods) have to be chosen and activated. For example, a scheduling model for a special ordering procedure (see section 5.1.3) may be used for materials with a high share in sales may be selected.

One needs to specify which of the data fields available in the software are to be used. Finally the forms necessary for the area to be supported are to be determined, and the desired reporting system needs to be set. These settings made within the software package are called *customizing*.

Some of the software firms and consulting companies that implement the software deploy pre-configured systems. They may include industry-specific or business type-specific adaptations which help to minimize the required efforts.

*Parameter settings* refer to how business objects represented in the software are dealt with. For example, one may specify which set of calculation procedures is to be used for different product lines, which material types will need to be order via which order process or how the minimum inventory level for the various materials will be specified.

After the software is implemented the master data inventory needs to be installed, or during a software change, existing data have to be transferred, supplemented and quality checked. Moreover, it is possible to implement individual, partially extensive supplements of the software package. The development environment in which the software product itself was developed in the first place is frequently used to do this.

After activating the application system, additional measures to fine-tune settings may be necessary in order to improve the technical performance of the software (e.g., response times) and the results of the deployed methods [Jäger et al. 93]. Training of the employees should begin well before the activation of the system.

## 6.2.3.3    Maintenance

Just as with any kind of software, software packages have to be maintained. On the one hand, the software producer supplies new versions to correct errors. On the other hand, newer releases may offer new functions and additional features. These features may be based on new knowledge from business research, as well as advancements in operating systems and systems software. The expenses caused by the introduction of a new version of a software package depend on the extent to which the software has been modified, as well as on the amount of changes to the data structure introduced with the new release. Generally, efficient maintenance of the software is only possible if all activities during the introduction phase of the project were clearly documented and updated. Sometimes, companies are forced to update to newer versions of the software. Otherwise, maintenance and support by the software vendor may expire due to contractual limitations.

Modifications of an established solution may be required if organizational changes within the firm are deemed appropriate. Moreover, new demands caused by products, as well as production processes may make new parameter settings necessary while the system is running.

# 6.3 Project Management

Project management tasks are planning and control of application systems development and implementation projects. Project organization includes further important aspects: tasks have to be assigned to the contributors and relationships concerning communication and cooperation have to be defined.

It is necessary to differentiate whether custom software will be developed by internal staff and, possibly, in cooperation with external contributors or if the focal concern is with the implementation and adaptation of packaged software. The underlying basic tasks are the same. What differs of course is the content that needs to be handled by the project management staff.

## 6.3.1 Project Organization

Development projects are typically managed by teams. The participating individuals take on differing tasks based on their qualifications and experience. Depending on the complexity of the project the assignment of tasks is done more or less formally.

In smaller projects (projects with comparatively few co-workers, a shorter duration time and a lower budget) it is common to encounter weakly formalized project organizations. This may imply that team members cooperate on an equal basis without defined roles. Informal discussion and coordination are used to distribute tasks and solve potential problems and conflicts. The advantage of this form of organizing is that no additional costs arise due to an excessive "bureaucracy". Flexibility and short decision-making cycles may persist.

For mid-sized and large development projects this form of project organization is of little usefulness: Agreements will not be adhered to (since there are no control mechanisms), conflicts remain unresolved or may be even unrecognized, etc. For this reason it is common to appoint a *project leader* in larger projects. This person is given the authority and responsibility to make far-reaching decisions. Moreover, the project leader coordinates the upcoming tasks and delegates these to team members. Members of the project team report the progress of their work, e.g., during status meetings.

The project leader does not always possess the sole decision-making authority and budget responsibility. Especially in very large projects it is common that the project leader operates under the control and advice of a *steering committee*.

To illustrate this situation, we assume that top management of a large industrial enterprise made the strategic decision to distribute its products via an Internet marketplace. For this reason a steering committee has been formed. Its membership is comprised of several specialists from the functional departments and the information systems group, but it also includes the chairman of the board, the board member in charge of sales, as well as the Chief Information Officer. During its first meeting, the steering committee appoints the project team leader and provides him or her with the necessary human and financial resources needed for a pilot survey. A few weeks later, the project manager presents the results of the pilot survey to the steering committee. Based on those findings the committee approves $ 1 million for the development of the technical concept and a prototype. During the course of the project the steering committee will provide direction-setting guidance and decisions (and especially will make available the necessary funds) in successive order. The project team leader remains obliged to report regularly and stay in frequent contact with the committee.

## 6.3.2 Project Planning and Control

Planning and control are the most important elements of project management. Usually the project leadership and the members of the steering committee are entrusted with these tasks.

Actual *project planning* consists of the following activities:

- Offices and authorities participating in the project are to be identified.

- Sub-tasks need to be coordinated with each other. In order to facilitate coordination, the project management has to specify methods and nominate co-workers who will take on the necessary tasks.

- Decision-making authorities of the participating contributors need to be clarified.

- For the introduction of software packages, the procedures to be utilized need to be specified. For custom software, e.g., the development environment needs to be determined.

- Task sequences for the development or implementation of the application system must be defined. One needs to examine how sub-tasks may be defined.

■  Dates on which the preliminary and final results of the software development or the software package implementation project are checked need to be specified (milestones).

One of the most important tasks of *project control* is *personnel leadership*. This pertains to technical aspects, but also to the perspective of employee motivation. In doing so the coordination of functional department and information systems interests is especially difficult (see section 7.2.3).

*Project control* examines if the tasks proposed during the planning phase were executed appropriately and if resources were used according to plans. Moreover, experiences are documented such that it becomes possible to build upon those in subsequent projects (controlling).

It has been reported that for many information systems projects the tasks to be accomplished could not be carried out in the planned for time and the estimated costs. Following we present therefore important procedures of project management that contribute to improved planning control.

An important planning tool for software projects is the *project structure plan* developed for the individual phases of the undertaking. Here one defines *task packets* that need to be accomplished such that the scope of performance of the application system may be delivered for the standard software or for the custom solution. In order to do this the tasks are partitioned into hierarchically ordered partial tasks. In the phase concept for software development we may detect broad task specifications already during the planning phase. The fine-tuning of these occurs during the definition phase. The task packets are monitored individually. The completion of a partial project for which appropriate controls are applied may be called a *milestone*. The necessary and appropriate assignment of personnel, needed computer and network resources, as well as additional software need to be determined.

For the various steps *project teams* need to be formed. One needs to consider that for the various task contents several specialists are needed whose availability is specified, i.e. limited in time. As liaisons between the information systems and functional department interests one often deploys *information systems coordinators* (see section 7.3.2).

Gantt diagrams or critical path tools are used to create project flow charts as aids for time scheduling. Finally one also has to estimate the costs of the project. For the control of the work project meetings are scheduled. For this, status reports are prepared which serve to examine adherence to technical demands and dates, as well as usage of resources and costs.

For scheduling and cost planning an *operating expenses estimate* needs to be developed. This expense estimate is being applied to all relevant resources separately. The appointed work force commands the largest proportion of this during development, as well as implementation projects. Consequently, these resources are planned for with special care.

Estimation procedures are fundamentally based on comparisons of projects or partial tasks with projects completed in the past. The quality of such procedures is mainly determined by the experience of the responsible experts [Boehm 00].

The scope of the software to be implemented, the necessary customization and modification are the basis for estimating the progression of the project during the standard software implementation. An orientation for this is given by the requirements specification list and technical specifications.

Software development projects are influenced by the size of the system (quantity), quality demands imposed on the solution to be developed, the project duration that needs to be adhered to, as well as the employee's productivity which may, e.g., be influenced by the software development tools used.

In the *Function Point Analysis* (cf., fig. 6.3.2/1) one uses for the operating expense estimation of software development projects a curve that describes the relationship between the development effort for an application system and its gradation on a point scale in so-called function points. Each effort is attributed to function points. These come about by weighting the functions of the new system, so-called business transactions, by using categories (e.g., number of data files to be processed) and by determining a point value also for qualitative factors (e.g., the complexity of the processing logic, integration with other application systems). For completed projects the derived sum of points and the actual effort is then transferred into a coordinate plane. The point value/effort curve may now be determined through a regression analysis. By using this curve it is possible to forecast the efforts needed for new projects in person months. In doing so the business transactions, as well as the qualitative factors are weighted and are converted into function points. It is then possible to read the corresponding value on the curve in number of months.

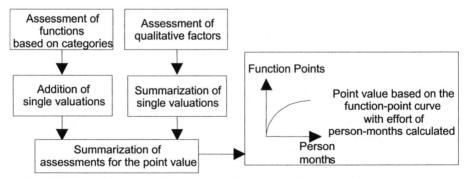

*Fig. 6.3.2/1*      *Operating Sequence in the Function Point Analysis*

In order to adapt the modus operandi of the function point analysis to, e.g., object-oriented development methods (see section 6.4.1.4), the function-oriented analysis is being substituted by an object-oriented one. Unfortunately, too little experience and no generally accepted rules of thumb exist in order to derive the effort curve [Sneed 96].

# 6.4 Tools for Project Implementation

A number of established means for supporting development projects have been created by both science and practice. Following we present the most important ones as an overview.

## 6.4.1 Modeling Techniques

In order to specify the technical demands in firms, as well as to document technical characteristics of the programs different descriptive techniques exist. Similarly, there are different methods to specify program modules. The following sections cover methods which have enjoyed a certain degree of success.

In business process modeling, operational processes are described. They are also used as well in order to document the handling of management tasks in software packages.

Data and functions modeling methods, as well as object modeling methods, are primarily used for the technical design in software development. Beyond this the documentation of the data model of software packages may, e.g., offer valuable assistance during the changeover of the data from old systems into new solutions.

### 6.4.1.1   Process Modeling

Before beginning with the selection, adaptation or development of a new application system one usually attempts to gain an overview of the relevant management processes in a firm. A *process* consists of a sequence of successively or partially parallel occurring functions and shows a defined starting point (trigger) and an endpoint. Such an operational flow is also called a business process or transaction.

*Process modeling* examines the purposes the individual functions fulfill within the entire transaction chain. In doing so one observes the sequence of the functions to be executed, their interfaces, as well as their attribution to respective organizational units. On the basis of process models one may, e.g., identify how it may be possible to reduce the through-put time of the process by improving the interfaces between the functions. Moreover, process analysis offers suggestions where eventually functions may be attributed to other organizational units in order to improve the work flow and process execution.

Target processes may serve as targets for organizational changes. From the information systems perspective, they contribute that an application system (that still needs to be chosen and to be developed) adequately supports the entire transaction chain as well. Some software vendors offer reference process models which allow potential buyers to evaluate in how far the processes implemented in the software correspond to the situation in the company. If one compares the reference processes to operational process models, one may realize how extensive an effort it would be to adapt the software to the operational realities. Or, in addition, one could identify which organizational changes would be necessary if the software was to be utilized without modification. Operational target processes may also be derived from such reference models.

In order to describe the application flow one often utilizes *event-driven process chains* [Scheer 94]. Fundamental elements of this presentational form are *events, functions* and their sequences. Events trigger the process, may alter the flow of functions and bring it to closure. Moreover, they are the result of functions. During the process flow variable branching is possible, e.g., when functions are carried out in a parallel fashion. Figure 6.4.1.1/1 depicts a simplified process segment of the procurement sector (see section 5.1.3). Starting with reaching the reorder level one needs to determine whether the material to be ordered is to be produced in-house or is to be procured externally, then the lot sizes need to be specified, as well as the procurement order or the manufacturing order need to be initiated.

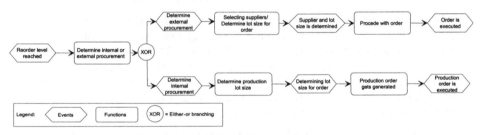

*Fig. 6.4.1.1/1    Excerpt from a Process Model*

Beyond this additional descriptive characteristics may be captured with this method:

- Organizational units responsible for the processing of functions and for their results

- Means or resources to be deployed during the processing of functions

- Data media on which the results are to be stored

## 6.4.1.2    Data Modeling

Data design starts with the description of the necessary data. Therefore the *relevant and logical data objects (e.g., customer and order data) and rela-*

*tions need to be derived* (see chapter 3) within an *abstraction process reflecting the firm's entrepreneurial reality*. The major tasks of this analysis are the collection of terms that each represents singular business-related circumstances. Then to clarify the meaning of these terms in order to depict them by the means of a formal descriptive language. Moreover, existing relationships between these data objects need to be determined. The resulting data model should reflect the operational reality as good as possible.

For that purpose we need to find largely interpretation- and redundancy-free agreements on semantics, i.e. the meaning of the terms. For example, one needs to define whether an item is an end-product, a material that still will go into an end-product or if it is both. This phase is therefore labeled *semantic or conceptual data modeling*. For this task the *entity-relationship model* established itself as a quasi-standard [Chen 76]. The conceptual data model is independent of how the data will be used. It may also be developed when the functions' details are unknown yet, as only the relevant data are being described and not how one is to work with them (see section 3.1.8).

The entity-relationship method is characterized by a clear definition and clear graphic depictions. With the *entity-relationship model* (ERM) static structures of data objects can be described and determined, as well as their relations. The fundamental elements of ERM are entities with their attributes, as well as the relationships between the individual entities.

*Entities* are real and abstract information objects with their independent, self-contained meaning. For example, an entity may be a customer, supplier or item, but also a department within a firm. In ERM we have to distinguish whether under entity we mean only singular information objects, e.g., a single, concrete customer, or if one denotes all entities of the same type, i.e. an entire class of "customer". In the latter case one refers to the *entity type*. An entity therefore is a singular, concrete specification of an entity type.

*Attributes* are characteristics of entity types. Their concrete specifications, the *attribute values*, describe the individual entity in greater detail. For example the entity type co-worker may be characterized by the attributes co-worker number, address, name, age and department. All entities of an entity class are described by the same attributes. They only differ through the values that the attributes take on in a concrete applied case. These values have to be within a value range that may also be labeled a domain.

Among actual entities certain *relations* may exist (e.g., customer A orders item X, customer B orders item Y). These can be classified among the entity types as abstract relations, i.e. *relationship types*. Basically we may encounter three forms of relationship types (see fig. 6.4.1.2/1) among entity types (see also section 3.1.8.1).

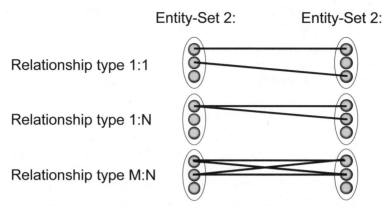

*Fig. 6.4.1.2/1    Forms of Relationships in the Entity-Relationship Model*

A 1:1 relation expresses that to each element of the first set belong to a maximum of one element of the second set and vice versa (e.g., each (singular) employee of a firm has each one employment contract). In a 1:N relation one entity of the first set may be assigned to no, one or several entities of the second set. Each element of the second set may only be assigned to at most one element of the first set (e.g.: No item, one or more items may belong to a class of goods; an item is assigned to no class of goods or is assigned to a class of goods). In the M:N relation each element of the first relation is related to none, one or more elements of the second set and vice-versa (e.g.: A certain customer orders no, one or more items and a certain item is ordered by no one, one or more customers). In an ERM we may encounter any number of entity and relationship types. Relationship types like entity types may also be characterized more specifically with attributes. In ERM one represents entity types through rectangles and relationship types through diamonds. The symbols are connected by undirected edges at which the complexity of the relationship type is noted. Figure 6.4.1.2/2 depicts simple entity-relationship models with different object and relationship types.

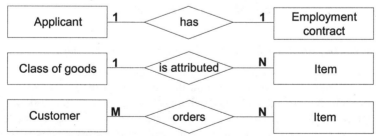

*Fig. 6.4.1.2/2    Examples for Relationship Types among Entity Types*

An ERM may be transformed into a relational database model rather easily. M:N relations are defined as separate tables via the foreign keys. 1:1, as well as 1:N models may be depicted through the complemented external key as an attribute in a relation (see section 3.1.9.1).

For the ERM numerous alternatives and expansions have been suggested. These render more precisely the complexity of relationship types or differentiate the specificity of relationship types [Stair/Reynolds 01].

### 6.4.1.3    Modeling of Functions

In a functions model, all relevant functions for an application system are being collected and structured. This, in turn, enables the observer to grasp the total system from a functional perspective. For that purpose one needs to depict the contents of the functions, as well as their interrelationships.

During the design of functions one mainly utilizes top-down procedures, i.e. starting with the problem statement the solution is being successively dissected into its individual parts. Such a process is complete when all relevant operational functions are identified and depicted. This dissection may go as far as to the level of a so-called *pseudo codes*, i.e. a precise description of individual processing steps (e.g., the sequence of screen pages). Figure 6.4.1.3/1 shows an example of a simple hierarchical diagram for a cost accounting system. It is subdivided into the modules cost element, cost center and cost object accounting. For the module cost center accounting a fine-tuned partitioning process is suggested.

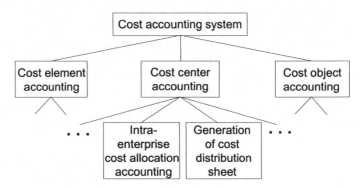

*Fig. 6.4.1.3/1    Functions Model of a System for Cost Accounting*

Alternatively, a bottom-up procedure, beginning with the modules on the lowest level, may be used. These modules are then joined to a total system. This approach is especially utilized when software packages are deployed in an integrated application system concept or if already existing software modules are to be reused. The bottom-up approach offers a lesser abstraction potential; it may affect adversely a clean structural design of the application system, as this, e.g., may be reflected negatively on the freedom for redundancy, as well as on maintenance where applicable.

Data flowcharts offer the graphical illustration of the information flow of an information systems application or of a function model. Using standardized symbols they illustrate:

■ which data are being read, processed and put out by a processing function (see section 3.1.2)

■ the thereby utilized data media

■ the direction of the information flow between the processing programs and the data media

■ the data type

Figure 6.4.1.3/2 illustrates the following circumstances in the notation of a data flowchart (for the classification of data see section 3.1.2).

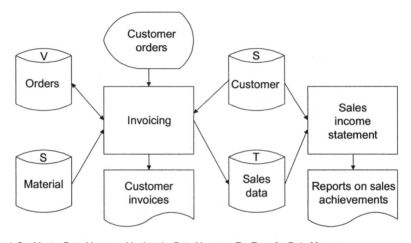

Legend: S = Master Data Memory  V = Interim Data Memory  T = Transfer Data Memory

*Fig. 6.4.1.3/2     Example of a Data Flow Plan*

Completed customer orders are sent via screen input to the program "invoicing". In order to read in the necessary data for issuing an invoice, e.g., customer name, address, item number, item price and ordered amount, the "invoicing" program accesses the master data "customer", "material" and the preliminary data "orders", issues the customer invoices and prints these. The sales data of the sold parts are stored in a transfer data memory. The program "sales income statement" generates reports about sales achievement for management. For this the information from the transfer data memory is read and will be linked to the customer master data.

## 6.4.1.4    Object Modeling

This procedure also utilizes the concepts of object-oriented programming (see section 2.2.1.2) and object-oriented data bases (see section 3.1.8.2) for the development process. This implies that data (here: attribute) and the functionality of their manipulation (here: methods) are brought together into a closed program unit, the object. Objects with the same characteristics and the same behavior, i.e. with the same methods and attributes, are combined to classes.

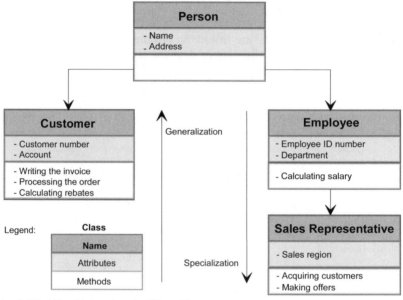

*Fig. 6.4.1.4/1    Example of a Class Concept*

So-called *inheritance relations* need to be defined to automatically pass along methods and attributes of a general class (upper class) to special classes (lower class). Figure 6.4.1.4/1 shows an example for a class hierarchy within which the characteristics of the class "person" are being bequeathed to the class "customer" and "co-worker".

A program sequence originates through the exchange of messages and information among the objects. They trigger an execution of a method at the receiving object, i.e. that this method is applied to attributes. To achieve the desired result the sender merely needs to know which messages he/she has to send. Knowledge about how the object internally operates is not necessary. A message is thus described by its name and through the specification of different parameters which determine its processing within the receiving object. If it is possible to send the same message to different receiver objects within a class hierarchy and thereby triggering differently implemented methods, we refer to this as polymorphism. If, e.g., the message "printing" within an invoicing program is being sent to the objects of the class "daily sales" and "monthly sales", and both belong to the upper class "sales", then the content of "printing" is different for both objects.

Object-oriented software development differentiates between a technical specification and an information systems-technical specification. In the technical specification the object classes, as well as their characteristics and their behavior are defined independently by information-systems-technical aspects. This corresponds to the specification of data and functions models during traditional procedures. Beyond that messages need to be determined between the object classes. With the information systems-technical specification (just as with traditional procedures) the structure of the application system, the processing logic, the platforms, as well as the lists are being shaped. The difference between both procedures is expressed as follows: The object-oriented applications development *largely takes the results of the technical specification over into the information systems-technical specification*. The objects and their structures remain unchanged. They are not being transformed as with the traditional procedure into elements of the information systems-technical concept, but are merely complemented by objects for the information systems-technical aspects [Bahrami 98].

Especially the Unified Modeling Language (UML) has achieved a wide distribution with object-oriented modeling [Fowler/Scott 00]. It supports linguistically, as well as visually the development process of the object-oriented requirements analysis up to implementation aspects for individual objects and their communication.

Object-oriented modeling reduces the "structural interruption" between the technical and the information systems-technical conception. This advantage compared to the traditional procedure is moreover reinforced in that both steps for application systems development utilize the same modeling components: classes (attributes and methods), messages, as well as inheritance.

Concept changes are easier to do, as the results of the technical and the information systems-technical model are stored together and they do not need to be carried over from one step into the next step. E.g., if one would like to modify the information systems concept, there is no need to return back to the technical conceptualization and affect changes there. With the object-oriented procedure only the affected objects have to be modified once.

Object-oriented modeling has the disadvantage that the integration with already existing software, e.g., during the joint use of relational databases, poses additional demands on the connected data modeling. Finally, object-oriented software building blocks lend themselves only for multiple usages, if also well documented, project-spanning class libraries are available. In such libraries software building blocks are deposited so that they may be utilized by other developers and don't have to be built from scratch. This requires additional organizational efforts.

## 6.4.2 Tools

*Software development tools* are defined as software systems that support one or more phases of the development process (as well as the herewith connected techniques and methods) from an information systems-technical perspective. In this context the term of Computer Aided Software Engineering (CASE) is also frequently encountered. The goal of the deployment of such tools is to ascertain the quality of the finished software products, as well as the efficiency of the development process.

*CASE-tools* that are mainly deployed in the early phases of software development are referred to as *Upper CASE-tools*. Essentially they support the modeling techniques described in section 6.3.1 by providing editors in particular for the graphic and textual capture and processing. Moreover, they check the formal correctness of models, carry out consistency checks between the partial models, control set of tasks, administer models and, finally deposit the results in a common database, the repository.

Even though the developers are being supported through these tools during the modeling, it should be emphasized that a "full automation" through CASE is not achievable. Certain aspects such as semantic definitions can principally not be depicted sufficiently in the software. Additional handicaps for the practical deployment are the high initial adaptation efforts, as well as the curtailment of the creativity of the co-workers. For the successful deployment of Upper CASE-Tools essential prerequisites are an appropriate preparatory training and a positive predisposition of the developers with regard to the structured, methodical proceeding in software projects.

CASE-tools that are deployed for the application in later phases of software development (especially during the implementation) are referred to as *Lower CASE-Tools*. Typical examples are *program editors* (for the compilation and processing of source code), *compilers* (for transforming the source code into an executable program), as well as the *debugger* in support of error search. In the practical world software development without such tools has become inconceivable in the meantime.

Apart from the already discussed programs there are software packages that support the phase-spanning management of projects, so-called project management software. With these it is possible, e.g., to generate Gantt charts, to manage resources, develop schedules and similar items.

## 6.4.3 Libraries

An important aspect of software development is the *reuse* of technical elements.

Within the framework of object-oriented programming so-called *class libraries* have been used for quite some time that offer standard interfaces, e.g., for the creation of platforms, for accessing the data system, the network or databases. The established libraries offer essentially technically-oriented interfaces.

On the technical level one may also fall back on proven concepts. Especially the so-called *reference models* (see section 6.4.1.1) have asserted themselves. They include, e.g., generic data and process models for various industries and firm sizes [Becker et al. 00]. Reference models are often the starting point when modeling and developing custom application systems. In comparison to the technically oriented libraries the deployment of reference models is, however, usually associated with greater adaptation expenditures.

More recently attempts have been made to bundle technical application logic in standardized and self-contained program building blocks. The fundamental idea is to combine such *components* (see section 6.1) using defined interfaces with each following the "building block principle" to larger application systems [Weske 99].

We also would like to point briefly to an additional interesting development. Programmers throughout the world are presently working free of charge on various software projects whose results are made available in source code at no cost on the Internet. Application systems developers may use these *open source programs* and adapt them to their own purposes and uses. Well-known examples with a high degree of distribution are the operating system Linux and Apache Web-Server that in terms of their qualitative characteristics stand fully on their own when compared to professional products. Whether this approach to software development will gain importance in other fields as well is difficult to judge today.

## 6.5    Literature for Chapter 6

Bahrami 98        Bahrami, A., Object Oriented Systems Development, McGraw-Hill/Irwin, Boston, 1998.

Becker et al. 00   Becker, J., Holten, R., Knackstedt, R., Schütte, R., Referenz-Informationsmodellierung, Bodendorf, F., Grauer,M. (eds.), Verbundtagung Wirtschaftsinformatik 2000, Aachen, Germany, pp. 86-109.

Boehm 00          Boehm, B.W., Software cost estimation with COCOMO II, Prentice Hall PTR, New York, 2000.

Chen 76           Chen, P. P., Entity-Relationship Model: Towards a Unified View of Data, ACM Transactions on Database Systems 1 (1976) 1, pp. 9-36.

Fowler/Scott 00    Fowler, M., Scott, H., UML distilled: a brief guide to the standard object modeling language, 2nd edition, Reading, Mass. 2000.

Jäger et al. 93    Jäger, E., Pietsch, M., Mertens, P., Die Auswahl zwischen alternativen Implementierungen von Geschäftsprozessen in einem Standardsoftwarepaket am Beispiel eines Kfz-Zulieferers, WIRTSCHAFTSINFORMATIK 35 (1993) 5, pp. 424-433.

Martin 02    Martin, R.C., Agile Software Development, Principles, Patterns, and Practices, Prentice Hall, 1st edition, New York, 2002.

Mellis/Stelzer 99    Mellis, W., Stelzer, D., Das Rätsel des prozessorientierten Softwarequalitätsmanagement, WIRTSCHAFTSINFORMATIK 41 (1999) 1, pp. 31-39.

Scheer 94    Scheer, A.-W., Business Process Enginieering – Reference Models for Industrial Companies, Berlin et al. 1994.

Sneed 96    Sneed, H. M., Schätzung der Entwicklungskosten von objektorientierter Software, Informatik-Spektrum 19 (1996) 3, pp. 133-140.

Stair/Reynolds 01    Stair, R.M., Reynolds, G.W., Principles of Information Systems, Course Technology; 5th edition, Boston, 2001.

Trauboth 96    Trauboth, H., Software-Qualitätssicherung: konstruktive und analytische Maßnahmen, 2nd edition., München et al. 1996.

Weske 99    Weske, M., Business Objekte: Konzepte, Architekturen, Standards, WIRTSCHAFTSINFORMATIK 41 (1999) 1, pp. 4-11.

# 7 Management Information Systems

From an enterprise-wide perspective we understand under the management of operational information systems the economical provisioning of all locations with digital device-generated information. This information is required to achieve the goals of the enterprise. Tasks related to this effort are referred to as (information technology-oriented) *information management*. Among others, it plays an important role how the enterprise positions itself within a competitive market.

The tasks of information management may be structured along a number of dimensions, e.g., along processes in planning, implementation and control or structurally in organization and scheduling. Following we address strategic planning of information systems and the organization of information systems as especially important scopes of duty and complement this discussion with additional strategic insights. In section 6.3 we already described with project management important questions of operational planning, control and organization.

Beyond this information management can provide information systems through which internal knowledge of the enterprise—e.g., technical details, decisions or project cycles—may be stored and be made available for future use. In doing so, tasks are taken on that are attributed to knowledge management. Important aids for knowledge management are content management systems that support workgroup-based creating and editing of contents of all kinds (see section 4.3.1.5).

## 7.1 Strategic Planning of Information Systems

The *strategic planning for information systems* is to determine the long-term measures of shaping the operational information supply. In order to achieve this goal, we may delimit three successive partial tasks: defining of an information systems strategy, the specification of an information systems architecture and finally the selection of information systems projects. The most important methodological tools in support of these three planning steps are presented below.

### 7.1.1 Defining an Information Systems Strategy

The general direction of a firm's strategic thrust of its information systems activities is determined by an *information systems strategy*. The general framework for this is provided by the firm's enterprise strategy. At the same time though it is also possible that impulses for changing of the enterprise strategy may be derived from the information systems strategy. Both aspects are addressed in the following sections.

### 7.1.1.1 Deriving the Information Systems Strategy from the Enterprise Strategy

Based on the purposes that an enterprise pursues with its deployment of information systems we may identify three strategic starting points. They are depicted in Figure 7.1.1.1/1.

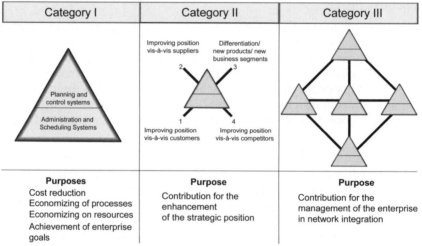

*Fig. 7.1.1.1/1   Starting Points for an Information Systems Strategy (based on [Mertens 00])*

Over time an enterprise may attempt to configure itself such that it evolves additionally from category I to category III.

■ *Category I:* The enterprise utilizes existing *potentials for cost* reduction. A manufacturing firm, e.g., invests in CAM (see section 5.1.5.9) in order to substitute expensive labor through automated manufacturing techniques. It would be possible to deploy in the sales area offer quotation systems (see section 5.1.2.1) in order to make sure that the responsible person in charge may be able to handle more offers than previously within the same time period. An integrated system for sales and production capacity planning may serve to scale and schedule the production line appropriately.

■ *Category II:* The enterprise attempts to improve or to hold its competitive position by offering the customer an output, performance or an added value through information technology by which it may differentiate itself substantially from the competition (cf., fig. 7.1.1.1/1). The intention is that the sales volume can be increased or that it will be possible to command a higher price than the competition.

*A PRACTICAL EXAMPLE*

*The Acme Credit Insurance Company offers its customers the possibility to utilize online-supported closing of an insurance policy for customer credits (commercial credit insurance). Compared to other insurance companies this company is able to offer a high level of customer service since most insurance decisions can be made online without any time delay. If the insurance company is unfamiliar with the company to be insured, the system delivers relevant data directly from information providers, judges and decides automatically through the use of expert systems [Schumann et al. 97].*

■ Category III: The company is managed within an *integrated network* such that it may adapt itself flexibly to changed environmental conditions. The integration in this may exist within a group of consolidated companies but also with other enterprises. Especially important here are the data exchange via communication networks, as well as the access to common databases. One form of information systems support for knowledge management are, e.g., "know-how" databases in which one may store and inquire very flexibly the knowledge and experiences the coworkers in the integrated network possess.

*A PRACTICAL EXAMPLE*

*First Regional Bank is an example for an enterprise organization that utilizes information systems to a very high degree. Aside from a central call center in Chicago there are others in Phoenix and Atlanta that are connected via communication networks. The calling bank customer does not know which location is presently helping him/her. For the central information systems applications First Regional Bank is also connected with a service computer center of a service delivery firm in Detroit. This organizational form enables the bank to mobilize additional service support if required.*

In the interorganziational area information systems enable different new cooperative forms, especially supply chain networks and so-called virtual organizations [Mowshowitz 02] which are outlined in section 5.4. Virtual organizations are order-related, interorganizational cooperations of legally independent partners with equal rights. The basis of this cooperation are agreements among the partners in terms of type and scope of that cooperation. Coordination within such networks occurs often through the support of information systems. For example, the coordination can base on electronic exchange of data or electronic markets for the internal distribution of orders.

*A PRACTICAL EXAMPLE*

*A subsidiary of Southern Real Estate Inc. that manages 140,000 apartments distributes electronically each day the received repair requests to about 400 participating craftsmen. All information for the order processing including the ac-*

*tivity confirmation and invoice dates are being exchanged via a communication network [Strohmeyer 92].*

## 7.1.1.2    Organizational Change through the Information Systems Strategy

Especially in information-intensive industries the mere derivation of information systems strategy from the enterprise strategy is not sufficient. Especially with the benefits of the interconnection through the Internet, it is becoming increasingly important to explore systematically also options of information systems for the enterprise strategy and, in doing so, to pay particular attention to the feedback loop. For this we recognize a market- and a resource-oriented perspective that complement each other.

The points of origin of a market-oriented perspective are the offered products and services. Information systems offer an enterprise opportunities to:

■ create entirely new markets,

■ develop business segments emerging from previous activities, as well as

■ support the development of new or the complementing of existing products.

A new market that establishes itself successively through networking are Internet auctions. The providers of such Internet auctions achieve their sales through commissions that they receive as a percentage of the auctioned goods, entry and participation fees of the buyers and sellers, advertising revenue from the Internet website, as well as the auctioning of goods that have been acquired directly to be sold through an auction.

An enterprise may create a new business segment when it combines a product or parts of it with the possibilities of information systems.

*A PRACTICAL EXAMPLE*

*The distribution of books through clubs is an established business segment for the Bloomingdales Group. This segment is complemented by the electronic book trade over the Internet through the online company bloomingdales.com. With the design of this form of distribution it was necessary to coordinate the products offered via the two distribution channels as much as possible, as well as to redefine the relevant business processes and support functions for the online book trade.*

Through the use of information systems products may be complemented, e.g., through customer service via communication networks. For instance, newspaper publishers enhance their classic print product though additional information, archives, discussion forums or via services for the administration of print subscriptions. Furthermore, many newspapers offer customized newsletters via e-mail which inform periodically about the selected topics. Even though this is an example of an enhancement of the classical print product, it may well be that in the future we assume that the concept of a newspaper also includes these described Internet services. Maybe we are witnessing a redefinition of the product 'newspaper' as we know it from the past.

A guiding framework for systematic checking of market-oriented options is advanced by Porter's [Porter 99] *analysis of competitive forces* network. Accordingly, there are five competitive forces of importance: the existing rivalry among established competitors, the threat of the achieved competitive positions through new market entrants and through substitute products, as well as the bargaining power of buyers and suppliers. According to Porter the strength of these competitive forces determine the earning potential. Basically, the earnings expectations get lower with increasing intensity of these competitive forces. Based on this approach figure 7.1.1.2/1 depicts an analysis of the market for search services. On the Internet search services make relevant addresses of information supply within a topic available. This occurs either through the input of a keyword or through navigating within a catalogue. With search services we recognize an important variant of portals (see section 5.1.3.2). Another variant are the so-called aggregators that not only deliver references for other Internet suppliers but they also offer their own content directly.

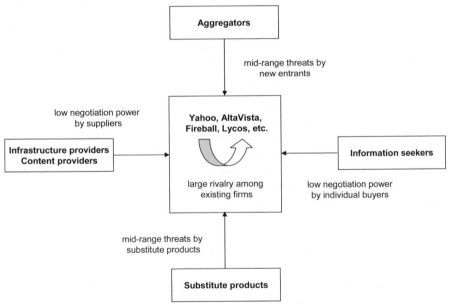

*Fig. 7.1.1.2/1   Market Analysis for a Search Service [Schumann/Hess 00, p. 34]*

It is important to consider the specific recurring patterns of information-based markets. Among these, particular attention is being paid to *network externalities* and also to *windowing* [Zerdick et al. 00]. Network externalities appear always when the user of a product realizes that for each individual user an added value or additional benefit occurs through networking (e.g., due to a communication network or off-the-shelf software). The main idea behind windowing is to distribute information via different distribution channels at different points in time and thereby maximizing one's total profit commensurate with the customers' willingness to pay.

Following the resource-oriented perspective one begins with the operational production factors and studies in particular individual activities. From this perspective, information systems offer possibilities to:

■ change the value added in an industry fundamentally,

■ connect individual value-added activities that are distributed within different enterprises among each other, as well as to

■ reshape and reconfigure individual value-added steps within an enterprise.

Information systems-based distribution channels lead to the disintermediation of traditional traders [Zerdick et al. 00]. Through this some established layers in the distribution chain (manufacturer ⇨ wholesaler ⇨ retailer ⇨ buyer) may disappear [Benjamin/Wigand 95; Wigand 97]. Even if full disintermediation does not occur, often the players at the distribution chain change entirely. For instance, airline reservations may be easily made on the Internet today. Payment occurs via a credit card with which one may also acquire the boarding pass via a ticketing machine at the airport (see section 5.2.5.2). Aside from this streamlined booking for the customer, the labor-intensive process of printing and delivering a paper ticket to the customer by the travel agent is beeing by-passed: If the customer chooses to work exclusively only with travel booking services or the airlines on the Internet, then he/she no longer would work with the traditional travel agent. We may assume that the traditional travel agent has been by-passed and is quite possibly disintermediated. Over 60 % of all airline travelers in the United States book their own airline tickets via the Internet today. Similarly, in the financial services industry, e.g., insurance companies, increasingly distribute their products over the Internet. Here as well the traditional sales force or independent insurance agents may be threatened by the above described trends toward disintermediation. Several insurance companies, e.g., Allstate Insurance Company, the largest home and car insurer in the United States, is pursuing such a policy aggressively.

Increasingly enterprises also take on the opportunity to create a network with their customers and suppliers. The online bookstore Amazon.com, among others, offers its customers buying tips. These are based on past buying choices and preferences which are compared to other buyers with similar interests. Large enterprises attempt to develop ever-tighter relationships and coordinate their business processes such that just-in-time deliveries, i.e. directly delivered onto the assembly lines, are possible. The overall implication is of course that procurement costs are drastically reduced (see section 5.1.3.2). Some companies, such as Dell Computer Company, have nearly perfected this supply chain challenge. Dell does not build a computer until the order has been placed and payment has been made or arranged for. It can then ship the computer to the customer within 48 hours. More recently, Dell—partially as a play on words—even talks about not just just-in-time delivery, but of just-in-the-nick-of-time delivery, suggesting yet tighter relationships with suppliers. Looking at the larger picture of these developments, one almost could assert that today's warehouses are a thing of the past and that "warehousing" occurs in an ideal sense in that the goods are "stored" while the trucks are moving on the roads to the factory, i.e. the need to warehouse has been eliminated. One should note though that this is a highly desirable ideal, but clearly not possible for all enterprises or in all application areas and industries.

*A PRACTICAL EXAMPLE*

*Bayer™ deploys a product catalogue on its Intranet that contains the goods offered by various suppliers. These present the necessary data sets such as pictures of products and information about the ability to deliver within a pre-specified format. Bayer employees select the respective products from the catalogue as needed and place these into a shopping basket. This basket is linked to an SAP order system that in turn transmits an EDI-based order to the supplier. The billing data exchange also occurs electronically [Fink 99].*

Beyond this, information systems also offer the opportunity to reshape and reconfigure the value added steps within an enterprise fundamentally. Under the superordinate concept *Business Process Redesign* these possibilities had been intensively discussed during the 1990ies. However, many enterprises never took full advantage of these possibilities.

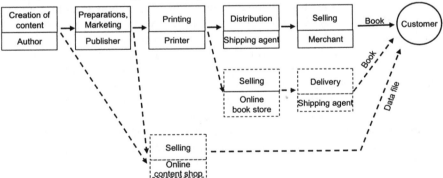

*Fig. 7.1.1.2/2    Simplified Value Chain Analysis for the Publishing Sector*

By using simplified *value chain analyses* [Porter 99] it is possible to identify configuration options systematically. Such an analysis attempts to dissect a firm or a line of business into its most important value-adding activities. A value-adding activity describes a technologically or physically delimiting activity. Figure 7.1.1.2/2 depicts a value added analysis that describes the changes to be expected within the fiction sector for online book traders and for the sale of digitized content.

## 7.1.2  Determining the Information Systems Architecture

The general direction and thrust of the information systems deployment of an enterprise is determined by its information systems strategy. An additional planning step helps us to substantiate this information systems strategy. For this purpose we build information systems architectures. Accordingly, this permits us to describe a longer-term development plan of the operational side of information systems—somewhat analogous to a building construction plan—at a relatively high aggregation level. Components of an information systems architecture are the information systems (IS) architecture and the information technology (IT) architecture.

The *IS architecture* focuses on the deployed application software. Its goal is to recognize gaps and redundant building blocks (programs or data) within the IS architecture of an enterprise. It is suggested that the enterprise model is a combination of the enterprise data and the enterprise functions model and deploys the known perspectives from data and function modeling.

Data integration, i.e. the utilization of common databases through different functional areas and application systems, requires the *design of conceptual data structures* at the enterprise level. Such enterprise-wide data structures constitute the *enterprise data model*. The goal of the enterprise data model is usually to clarify the interrelationships among the technical tasks in an enterprise and the necessary data. This procedure allows to identify those data which can be used for multiple tasks. Moreover, at the same time a prerequisite is created for the data-oriented integration of different system components. With the *enterprise function model* an overview is offered suggesting which functions need to be carried out in the enterprise and how they may be structured. The model permits to match which task areas in the enterprise are already supported by information systems and where additional needs for information systems deployment exist. The enterprise function model describes in which context the data specified in the enterprise data model should be used. One also refers to the *comprehensive enterprise model* when an integrated view of data and functions is taken. Figure 7.1.2/1 shows excerpts of an enterprise data model and a comprehensive enterprise model.

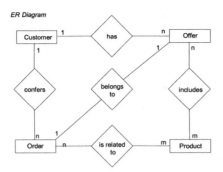

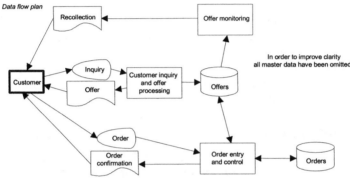

*Fig. 7.1.2/1     Excerpts from Enterprise Models*

An alternative to data and function orientation is business process orientation. During the development of an enterprise process model usually those business processes (core processes) are focused whose implementation have a sustainable effect on the competitive position of the enterprise. For this purpose, e.g., three classes of processes may be differentiated [Hess 96, p. 166]:

■ *Performance processes* include tasks needed for the marketing of products and services and in doing so utilize existing capabilities.

■ *Support processes* are all tasks that are necessary for the set-up and maintenance of resources. They create capabilities that may be utilized by performance processes.

■ *Management processes* consist of comprehensive planning tasks within an enterprise. Management processes in particular guide utilization and build-up of capabilities.

The already presented modeling techniques in section 6.4.1 are well suited to generate an enterprise process model.

Aside from determining the IS architecture, the *IT architecture* also has to be specified. Analogous to the IS architecture the IT architecture also needs to be defined in terms of long-range, relevant decisions shaping the technical infrastructure at a high-level aggregation. In doing so the focus here is with hardware, system-related software, as well as technical network questions. Within the framework of the IT architecture we need to determine which operating systems and DBS are to be deployed and on which basis protocols and typologies bring about the networking of computers. TCP/IP protocols have asserted themselves (see section 2.5.1) as a standard for the networking of heterogeneous computers.

Additionally, we would like to point out two current questions for the determination of information systems architectures. Fundamental changes in the technologies and in the entrepreneurial challenges make very long-term planning problematic. For this reason one tends to concentrate with the definition of IT architectures more and more on few, but fundamental aspects as compared to very detailed modeling. Moreover, one may recognize that individual enterprises are less and less capable to determine their information systems architectures autonomously. Through cooperation, as, e.g., with supply chain networks (see section 5.4) the elbowroom of an individual enterprise is being limited. As a consequence of these developments the definition of standards is becoming increasingly important.

## 7.1.3  The Selection of Information Systems Projects

The implementation and transformation of an information systems architecture occurs through projects. Since the resources for such projects are scarce, one needs to prioritize. Such a prioritization should orient itself on the following criteria:

1.  Influence on *achieving enterprise goals*

2.  *Effects* on technical tasks

3.  *Technical feasibility for implementation*

4.  *Cost effectiveness* of the information systems project

Following we discuss how information systems projects may be assessed fundamentally. Moreover, we will show how cost effectiveness may be judged.

### 7.1.3.1    Information Systems Project Portfolios

A frequent procedure to evaluate multiple information systems projects at the same time is *portfolio analysis*. Individual information systems solutions are placed into a matrix. This positioning occurs through a subjective evaluation, e.g., by the employees of the functional department and information systems department. Portfolios may be used to evaluate enterprise goals and functional areas. But as well, portfolios may support technical and implementation-related judgments. Figure 7.1.3.1/1 shows two examples.

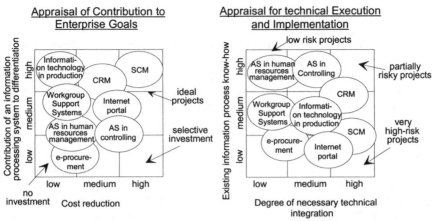

*Fig. 7.1.3.1/1   Portfolio Analyses for the Appraising of Information Systems Projects*

The enterprise goal-related portfolio uses two axes: the potential competitive differentiation and the potential for cost reduction. On the other hand, a technological evaluation takes place, using a portfolio which depicts the existing IS know-how on one dimension and the necessary integration with existing information systems on the other.

A high degree of integration suggests larger difficulties during the implementation. Based on this judgment one may choose information system projects that, on the one hand, contribute to the achievement of the business objectives and, on the other hand, allow the distribution of risk while the technical implementation takes place. In the example mentioned above the information systems project for customer relationship management (CRM, see section 5.1.2.4) would have positive effects on enterprise goals. Due to the high degree of integration technical difficulties may be expected for this application system. The risk of implementation for a project portfolio may be acceptable, if the projects are distributed well within the technologically-oriented matrix, i.e. over several fields within that matrix.

## 7.1.3.2   Analyses of the Profitability of Information Systems

Profitability analyses, usually assigned to *information systems controlling*, serve during the planning for information systems solutions as decision aids whether or not an application idea should be implemented. During ongoing information systems operations it is possible to check whether or not information systems have the planned effects.

For planning purposes cost benefit analyses have to be carried out at the beginning of the information system project. For particular application systems they then may be detailed parallel to the implementation process. When we look at the analysis of cost benefits we need to consider *directly, monetar-*

*ily appraisable effects,* as well as the *strategic character* of an application system as a non-monetary factor (see section 7.1.1).

Through monetary values it is possible to conduct investment calculations. Static analyses, i.e. analyses based only on one period, may only provide initial suggestions. More suitable are multi-period *dynamic investment calculations*, since usually costs as well as efficiencies for information system change over time. At the beginning development costs are higher (e.g., for programming or the implementation of software packages), whereas in later periods the maintenance costs of systems dominate. In addition one needs to differentiate between one-time costs for planning, system acquisition and implementation preparation, as well as recurring costs of general operating systems. The costs and benefits are often subject to sizable fluctuations according to the life cycle of an information system project. At the beginning users often need to be educated. Usually economies of scale are not observable until operations have become routine.

Estimating benefits, especially for integrated systems, is usually more difficult than cost determination. Aside from direct monetarily quantifiable benefits (e.g., an automated operational function leads to personnel cost savings), we recognize results that are monetarily not directly quantifiable (e.g., increasing punctuality in the shipping department), as well as qualitative or strategic effects (e.g., the increase in product variability from which the customer may choose). The intention is to qualitatively determine as many factors as possible. Aside from monetary results one may also use time units (for the comparison of activities with and without information systems) and similar measurements. Such indirect measures subsequently must be converted into monetary units. Qualitative effects may be added up to an *argument list, as shown* in figure 7.2.1/1.

Beyond this, benefits may appear in other places or functions than the immediate usage site of the system, especially while using integrated solutions. As an example we would like to refer to an information systems application of an incoming goods inspection control (see section 5.1.3.4) that supports a sampling inspection. Since this system is based on sophisticated statistical procedures, it is possible to recognize this way a larger number of qualitatively sub-standard deliveries than it would be possible with individual sampling inspection decisions. As a consequence we also encounter fewer difficulties in manufacturing. This in turn leads to the lowering of costs for subsequently needed improvement and the rejection of defective goods during production. Thus the inspection system used in the goods receiving area leads to savings in the production area.

In order to capture the efficiency effects precisely, one may ascertain commensurate estimates, e.g., through the use of *cause-effect chains*. For example, shortened item storage times managed by an information system result in reduced warehouse inventory. Thus we are less tied to capital expenditures, encounter a smaller inventory risk and enjoy a smaller need for storage. Moreover, we state that under certain assumptions shorter cycle times that lead to an improved market position of the enterprise may increase the sales volume and thus result in greater marginal income.

On the basis of identified benefits and costs the profitability analysis takes three steps:

1. The appropriate measures for the information system application are ascertained and potential effects are identified.

2. The effects need to be valued.

3. The *net benefits* are calculated by the subtraction of the costs from the benefits (gross).

In the second step one has to decide whether and how the identified changes may be quantified. Since many of the used data are afflicted with uncertainty and imprecision, one often uses not only one measure, but estimates optimistic, probabilistic and pessimistic values or considers ranges in the calculations. The quantitative result is mostly supplemented by an *argument list (or arguments balance sheet)* in which the qualitative effects of the new solutions are specified. Both aspects are then considered for a comprehensive judgment.

Figure 7.1.3.2/1 shows a strongly simplified example of this calculation considering the deployment of a CAD system. The *net present value* serves as a criterion for judgment. Cycle time and quality improvements during the drawing creation are supposed to lead to increased success in that market and may thus result in marginal income increases. In order to look at the effects of subsequent time periods, the following five years are considered and the corresponding results are discounted with an interest rate of 6 %.

Often it may be important to question the effects for the enterprise, if the information systems investment would not be made. There is the danger that other enterprises that do invest into this information systems solution will gain a competitive advantage and that thus one's own enterprise will have to face sale losses. For example, during the introduction of reservation systems in travel agencies it was observable that airline companies that were not participating in this reservation system lost market share.

| Financial consequence Dollar    year | 0 | 1 | 2 | 3 | 4 |
|---|---|---|---|---|---|
| Outlay through the CAD System | | | | | |
|   Amount to be invested | - 120,000 | | | | |
|   Maintenance / ongoing costs | | - 15,000 | - 15,000 | - 15,000 | - 15,000 |
| Direct Effects | | | | | |
|   Personnel costs savings through faster drawing development | | 30,000 | 65,000 | 65,000 | 65,000 |
| Indirect effects in other areas | | | | | |
|   Reduced Capital Costs through increased usage of standardized components | | 5,000 | 10,000 | 15,000 | 15,000 |
|   Lower costs in job preparation | | 10,000 | 15,000 | 20,000 | 20,000 |
| Increase in marginal income contribution | | 5,000 | 15,000 | 20,000 | 15,000 |
| Net Efficiency (NE$_y$) | - 120,000 | 35,000 | 90,000 | 105,000 | 100,000 |
| Calculated interest rate | 6 % | | | | |
| Net present value | 160,488 | | | | |

Formula for Capital Value

$$NPV = \sum_{y=0}^{n} \frac{NE_y}{(1+i)^y}$$

NPV: Net present value in $
i: calculated interest rate
NE$_y$: Net efficiency in $/year
y: year
n: Length period under consideration in years

*Fig. 7.1.3.2/1    Evaluation of a CAD Deployment*

For the final selection of information system projects the selected applications need to be compared to available resources for actual realization and implementation. Therefore the available budget, human resources, computer and network resources, available for the program creation, need to be considered.

Aside from cost effectiveness considerations for individual information systems or projects, respectively, information systems controlling also encompasses the differentiated planning and control of the necessary resources for the going concern, such as human resources or hardware. Likewise the performance to be delivered by the information systems department needs to be planned and evaluated using quantitative and qualitative aspects (e.g., response times for dialogue applications, scope of the new programs, programmer productivity) [Reichmann 97]. In order to evaluate application development and usage one utilizes, e.g., a performance measurement system.

Sometimes a return on capital investment measure is utilized that the enterprise minimally wants to achieve with its information systems investments.

## 7.2 Organization of Information Systems

An additional task, as part of managing operational information systems, is its *organizational design*. On the one hand one needs to consider the individual functions, as well as the people responsible and the organization units (e.g., departments) which are integrated in the operational organization structure (*structural organization*). On the other hand, one has to determine how the processes have to run within information systems (operating sequence or *process organization*). When designing the organization's structure three questions are of particular interest:

1. Which tasks have to be taken care of in one's own enterprise, which ones will be taken care of externally?
2. Where is an information system organizationally located and to which unit does it report?
3. How is information systems structured itself?

Following these three questions are addressed briefly.

### 7.2.1 Internal Production or External Procurement of Information Systems Performances

Every enterprise has to decide if the required information system performance should be generated internally or obtained externally. Such decisions may be made for the entire information system unit, the use of a service and computer center and a network provider, the entire delivery of application systems and the internal or external procurement of individual programs.

The decision to draw information systems performance from external sources which were internally delivered previously is referred to as *outsourcing*. The performance of external firms is used in some areas (e.g., in technical network management) when highly specialized know-how is required. Moreover, the performance of one's own specialists is often more expensive or even qualitatively worse than externally obtained products. For example, the Xerox Corporation outsourced its entire information system area to EDS, including the staff, for a contracted ten year period. The outsourcing of a major area may also be associated with disadvantages such as the loss of the firm's own know-how in the information system area and a dependence on the chosen outsourcing partner. Moreover, it is questionable if, after ten years, the firm would still be in the position to take the information system area back or start a new one, even if it wanted to. Arguments for and against the external procurement of information system services are presented in figure 7.2.1/1.

| Internal Production | Arguments for | External Procurement |
|---|---|---|

- ○ Existing entrepreneurial and information systems know-how may be utilized for the production of goods and services

- ○ Information systems tasks of strategic importance may lend themselves to establish barriers for the competition

- ○ No irreversible dependencies from other firms

- ○ High degree of proximity may lead to good acceptance in various sectors of business

- ○ No costs will occur for the coordination of external suppliers

- ○ Concentration on the core business of the enterprise

- ○ Access to internal unavailable know-how

- ○ Quicker availability of capacities

- ○ Even utilization of human resources, e.g., for tasks with low frequency of occurrence

- ○ Cost reduction for individual tasks

- ○ Avoidance of human resources recruitment for tasks that occur only temporarily and occasionally

*Fig. 7.2.1/1      Selected Arguments for Internal vs. External Procurement (based on [Mertens/Knolmayer 98, p. 34])*

The decision between internal vs. external procurement of an information systems service is determined by its strategic importance and through the performance delivery capability of the firm for the considered tasks. Firms that generally have problems with the quality of their information systems service and with the deployed state-of-the-art technology are more likely to outsource than those firms that have better skills in the information systems area. With standardized information system services that are not especially tailored to the organization, outsourcing becomes possible. In contrast information system services of strategic importance for the enterprise are less suitable for outsourcing.

More recently a special form of outsourcing, i.e. Application Service Providing (ASP), is increasingly gaining importance [Tamm/Günther 00], In this connection the user has access to the available application system via an Internet browser, which means more flexibility. In addition the applications are generally not tailored to the individual perceptions of the user.

## 7.2.2 Placement of Information Systems in the Business Organization

Basically operational information systems may be placed in two different ways in the organization of the business: as a staff unit of executive management or as a functional area. Figure 7.2.2/1 shows alternatives for this purpose.

The organizational placement as a *staff unit* emphasizes the *service nature of information systems* within an enterprise. In the example it is directly attached to the executive management (①).

When information systems are of particularly high importance for the enterprise as a whole or when, e.g., information system services are being offered on the external market, an integration under the executive management as a *functional area* comes into consideration (②). One must recognize though that the intraorganizational service role diminishes in this alternative [Mertens/Knolmayer 98, pp. 52].

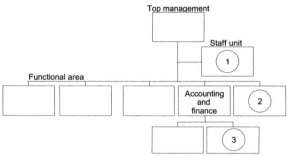

*Fig. 7.2.2/1*    *Organizational Fit and Integration of Information Systems Departments*

The attachment to particular departments is largely due to historical and evolutionary reasons. During the times when accounting managed data processing activities primarily, information systems was often attached to the finance and accounting area (③).

Moreover, within a matrix organization information systems may be structured like a cross-sectional function, i.e. it plays a role in all functional areas [Mertens/Knolmayer 98, p. 54].

Enterprises with a distinct decentralized information system possess aside from the centralized information system department (largely taking on coordination tasks), *decentralized groups* that are responsible for specific information system services in functional and business areas.

Some enterprises have outsourced information systems or parts of them, which have become legally independent firms. The enterprises utilize the information system services of the newly created firms or maybe even sell their information system operations to them. The advantages of working with those autonomous firms may include, e.g., the possibilities to sell in-house designed and programmed software to third parties more easily, as well as the ability to check the efficiency of the firm's own information systems performance processes through a comparison of quality and price within the open market. Furthermore, the outsourced firm is able to pursue its own independent salary and employment strategies and to offer more flexible working hours than this might have been possible before, within the rigid structures of the parent company. The latter point may be a considerable guidance for software development projects.

Large software development departments have relocated fields of functions to countries where the wages are lower (offshore outsourcing). The necessary coordination activities are then done via telecommunications services. Bangalore in India, e.g., has become well known as a center for those activities within the software industry.

## 7.2.3 Internal Organization of the Information System Area

Tasks of the information system department are, e.g., the strategic information system planning, as well as providing methodological support for the transformation of business processes, the development of its own application systems and the introduction of software package, running the computer center and networks, as well as user support services.

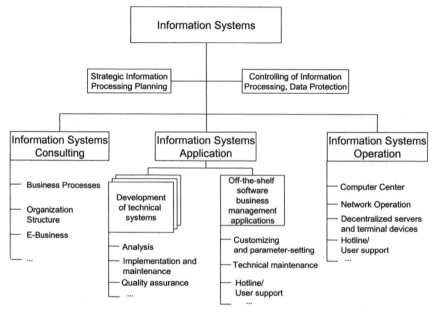

*Fig. 7.2.3/1    An Example of the Organization of an Information Systems Department*

Often the design of the general flow of operational processes and the development of electronic business (e-business) are assigned to information system departments, as well as data privacy and protection and the controlling efforts of information systems. A structuring of the information systems department could therefore be based on these tasks and functions. How detailed an enterprise subdivides its information systems department depends, among other considerations, on the size of the enterprise. A large enterprise may cover all functions, a mid-sized firm, however, may only emphasize the computing center and networks, as well as software package support. Figure

7.2.3/1 shows an example of an organization structure for an information system department. In the practical world many alternatives exist.

Individual functions themselves may be structured based on different criteria. Within the area of application systems development one differentiates, e.g., often according to functional aspects (e.g., application systems in the sectors human resources, accounting, materials administration, production) or departments of a company. This is influenced, among others, by the cooperation between the information systems departments and the functional departments during the application system development. In this context figure 7.2.3/2 shows five design configurations. Enterprises that usually deploy software packages also employ appropriate maintenance and support service. Employees in functional departments often develop smaller application systems on their local systems on their own. For this purpose they may use spreadsheet programs, PC databases or HTML editors (see section 2.2.2.1.1). This may lead to increased effectiveness and flexibility in the functional areas.

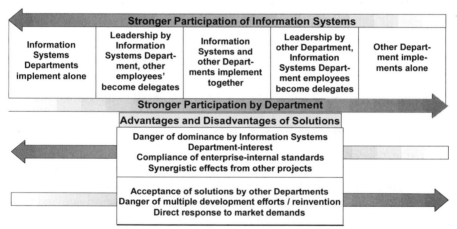

*Fig. 7.2.3/2    Participatory Models for Software Development Projects (based on [Mertens/Knolmayer 98, p. 87])*

Considering the so-called *individual data processing* one may notice that the complexity and also the development efforts of the problems addressed by the functional departments are limited (or must remain limited, respectively). The data volumes are usually low, often insular solutions are created and data protection and security aspects are neglected. In spite of the use of simple programs individual data processing may only be practiced effectively, if the user also receives sufficient support. Therefore enterprises created *User Support Centers* that support the user with his/her information system tasks, as well as the selection of suitable hardware and software. Enterprise-wide networks enable even firms with distributed locations to offer these services in a centralized fashion.

# 7.3 Additional Management Aspects

Aside from strategic planning and the organization effort itself information systems-related legal aspects and also professional career roles play important parts in the practical world. Following both perspectives are addressed briefly.

## 7.3.1 Legal Aspects of Information Systems

When deploying information systems different legal regulations need to be considered. A few selected issues are identified here.

- Protection of personal data
- Signature laws pertaining to the identify of individuals
- Computer crime (theft and the willful destruction of information system installations, computers and program via computer and network use)
- Infringement of copyright laws for computer programs
- Form of contract, such as with procurement and maintenance contracts for computer hardware
- Product liability for damage occurring through computer hardware and software

Following we address a few selected areas pertaining to the United States.

### 7.3.1.1 Data Protection and Data Privacy

The rising concern over the acquisition, use and dissemination of personal data, Internet privacy, the Safe Harbor Program, conflicts and disagreement with parallel European Union (EU) efforts and many other privacy issues have created considerable debate and frictions among those private and public sector entities involved in these topics. In the United States data protection and data privacy is increasingly making front page news and are the subject of considerable legislative activity at the state and federal levels.

In the United States the Supreme Court first recognized a right of *personal privacy* in 1891, which has at various times since been related to the First, Fourth, Fifth, and Fourteenth Amendments to the US Constitution. This implied *constitutional protection for privacy* has been extended mainly to limit government regulation which impinges on an individual's right to autonomous decision making on certain issues, and secondly to prevent the disclosure of personal matters. Privacy interests of employees are also affected at the federal and state level by constitutions, statutes, court decisions, administrative bodies, arbitration decisions under contracts, and common law causes of action.

The US Constitution may protect employee privacy in a range of situations. The collection and disclosure of information by employers may impinge on the First Amendment protection related to freedom of religion, speech, press, assembly and petition. The Fourth Amendment right to protection from unreasonable search and seizures is also relevant [Dixon 95]. The Fourteenth Amendment gives employees a right to a fair hearing and due process. Public sector employees have greater protections because these constitutional provisions relate to protection from "state action", although in some situations these rights may be applicable to some private sector employees.

Common law causes of action based on tort law may also provide employees with privacy protection. While tort theory varies among states, five major common law actions have been developed: invasion of privacy, intrusion upon seclusion, publicity given to one's private life, publicity placing a person in a false light, and intentional infliction of emotional distress. Unreasonable surveillance has been found to constitute both an invasion of privacy and an intrusion upon seclusion.

The *Privacy Act of 1974* applies to files generated by electronic monitoring in public agencies at the federal level. Digital recordings are included in this definition as a record is defined as any item or information which contains something which can particularly identify an individual. A limited range of data protection principles applies to the collection, storage and use of the recordings. Federal privacy laws are relatively weak and when it comes to law enforcement, typically state-level privacy laws have sufficient and better legal rigor to affect prosecution. An additional dilemma in the United States is that data protection and privacy legislation weaves in and out of many bodies of law at the federal and state levels. These topics may be addressed under many, sometimes merely remotely related, laws, legal applications and considerably varying contexts, as well as degrees of emphasis. In part this situation is due to historical reasons, i.e. law developed in the United States modeled after English law which is based on case law (with the exception of the State of Louisiana in which Napoleonic law prevails). As a consequence, legal interpretations are made on previous cases, but it would be impossible to identify, e.g., a body of law uniformly and categorically addressing 'data protection'.

The *EU Directive on Data Privacy* has had a significant impact on the United States. A considerable debate evolved in that the directive states that EU countries may not transfer personal data to countries that do not have 'adequate' privacy and data protection.

The United States does not have strict data privacy laws addressing all of the addressed principles in the EU Directive. The prevailing regulatory method in the United States relies instead on *self-regulation*. In turn, the EU deemed that the United States does not have "adequate privacy protection."

In addition to ongoing hearings and workshops involving various government agencies, the US Department of Commerce began working with the EU to develop the concept of *safe harbors*. The 'safe harbors' are a set of common principles providing a specified standard of 'adequate privacy protection' pertaining to the transfer of information from participating countries to the United States. The European Commission ruled on July 27, 2000 that the Safe Harbor Privacy Principles submitted by the US provided adequate protection.

According to the US Department of Commerce, "the safe harbor eliminates the need for prior approval to begin data transfers." Firms deciding to participate in the Safe Harbor Program must comply with the Safe Harbor requirements and must publicly declare their adherence. It should be noted though that participation in the program is entirely voluntary for US firms. Such adherence is expressed in that the participating firm self-certifies annually to the US Department of Commerce in writing that it agrees to adhere to the Safe Harbor Program's requirements. Moreover, the firm must also state in its published privacy policy statement that it adheres to the Safe Harbor Program. The US Department of Commerce, in turn, maintains a list of all firms filing self-certification letters. Both the list and self-certification letters are publicly available.

The Safe Harbor Program is comprised of seven principles:

1. *Notice:* Notice must be given about what is collected, how it is collected, its purpose and disclosure to third parties, as well as choices and means the firm offers for limiting use and disclosure.

2. *Choice:* Choice or opting-out allows individuals to choose not to have their information or data used or disclosed to a third party or to be used for a purpose incompatible with the purpose for which it was originally collected.

3. *Onward Transfer:* Transfer only to third parties who also comply with the Safe Harbor principles.

4. *Access:* Provide individuals with access to their personal information and allow them to correct, amend, or delete that information where it is inaccurate (except when the burden or expense of providing access would be disproportionate to the risks to the individual's privacy or where the rights of persons other than the individual would be violated).

5. *Security:* Reasonable precautions must be taken to protect information from loss, misuse and unauthorized access, disclosure, alteration and destruction.

6.  *Data Integrity:* Use data consistent with the purpose disclosed to the individual.

7.  *Enforcement:* Must have mechanisms for assuring compliance. Specifically, there must be:

    a)  Readily available and affordable independent mechanisms for the hearing of complaints by individuals and awarding of damages where law or private sector initiatives provide them.

    b)  Procedures for verifying that the commitments companies make to adhere to the Safe Harbor Principles have been implemented, and

    c)  Obligations to remedy problems arising out of a failure to comply with the principles.

    In addition, sanctions must be sufficiently vigorous to ensure compliance.

Lastly, with the ever-increasing importance of the Internet and WWW, a few words on *Internet privacy* are warranted. The collection and use of personal information over the Internet has been highly publicized recently. There has been a growing amount of legislative activity surrounding this topic. Numerous states have introduced legislation creating special task forces and committees to further investigate these privacy issues and concerns. Other lawmakers have, e.g., introduced legislation to prohibit email or Internet service providers from disclosing the personally-identifiable information about their subscribers (including e-mail addresses) without notice and consent.

In spite of these efforts, and admittedly only a few that are directly germane to our discussion have been addressed here, privacy and data protection is in the United States still a work in progress.

## 7.3.1.2 Authentication

Providing proof that a person really is the one he/she claims to be is called *authentication*. Given the worldwide activities possible on the Internet and its many forms of data exchange we quickly realize that the definite attribution of data or of an electronic signature especially with open user groups is of great importance (e.g., with money transactions).

An electronic signature is data in electronic form that are attached to or logically associated with other electronic data and which serve as a method of authentication. Moreover, it is a signature uniquely linked to the signatory, which is capable of identifying the signatory, can be maintained under his/her sole control and is linked to the data to which it relates such that subsequent change is detectable.

A signature in digital form attached to data should meet four requirements:

1. It is uniquely linked to the signatory

2. It is capable of identifying the signatory

3. It is created using means that the signatory can maintain under his/her sole control

4. It is linked to the data to which it relates in such a manner that is revealed if the data have subsequently been altered

As the Internet has no borders, it is essential for vendors and users alike to observe international common standards, to provide consistent services worldwide and to comply with various regional criteria. Although our space is limited to address in detail technological specifics pertaining to authentication, we would like to mention that Common Criteria have been developed by *public key infrastructure* (PKI) providers. The developed set of common criteria represents the outcome of a series of efforts by government organizations from the United States, Canada, France, Germany, and the United Kingdom among others to develop criteria for evaluation of information technology (IT) security that are broadly useful within the international community.

The Common Criteria is an International Organization for Standardization (ISO) recognized evaluation process, developed by a collaboration of industry and government agencies like the National Security Agency (NSA) in the United States, and others around the world.

### 7.3.1.3    Additional Legal Considerations

Copyright laws are to protect the creator of a software from unauthorized use, dissemination or further development of programs (e.g., bootlegged or pirated copies). Programs are then protected when they originate from an author and when they are the result of his/her own intellectual creation.

If a software package is acquired or if a software house develops individually for the customer a program, appropriate contracts are signed. When software is purchased, underlying this step is a *purchase contract* with which the buyer acquires a warranty usually for at least six months that the software's promised product characteristics are assured.

If a *contract for work and services* for custom programming has been signed, the customer usually accepts the performance of work and provided services, assuming this was accomplished satisfactorily, with a completion protocol or certificate. External consulting contracts, e.g., for the information systems organization, is usually handled via *service delivery contracts*. In those the fee for the activities is specified.

## 7.3.2 Professional Career Roles in Information Systems

The nature of work determines how information systems employees may be classified [Dostal 99, pp. 192]. One may identify:

- Core job
- Mixed job
- Fringe job

*Core jobs* are usually those in which the computer-related work is dominant. Most individuals belonging to this group are in the operational information systems department, with information systems service firms, with hardware and software vendors, as well as Internet firms. Especially in the latter professional area the roles and position descriptions change quickly.

One may identify four broad areas that are characterized by varying task emphases:

- Software development, implementation and maintenance
- Management of information systems
- Design of technical solutions (business processes, e-business models, etc.)
- Operation of information systems

In software development, implementation and maintenance one may, e.g., identify the following professions:

1. *Systems Analyst* or *Information Systems Organizer* examine the current, actual condition of existing systems and determine which new demands or changed processes in information systems applications should be implemented. They often develop solution proposals or target concepts.

2. *Programmers* develop on the basis of predetermined specifications information systems applications or modules for individual application systems. Based on qualification and experience of the programmer one often differentiates between *junior and senior programmer*.

3. *Multimedia Designer* and *Manager* have the duty to prepare information in a user-friendly way for external display of an organization in order to present it on the World Wide Web and to integrate the services of the Internet with business processes.

4. The *Consultant for software packages* supports enterprises during the design of business processes, the customization and the establishing of parameters, as well as the maintenance of software packages.

As selected professional positions in information systems management we would like to identify:

1.  The *Information Manager* (*Chief Information Officer, CIO*) is responsible for strategic planning, preparation and execution of tasks within the information systems area within the enterprise, but also in interorganizational settings (see section 7.1).

2.  *Information Systems Auditors* assure security, accuracy and orderliness of information systems applications.

Important for business information systems is a professional image in whose core is the development of technical, information systems-based solutions. Some modern expressions of this are:

1.  Coordinators for supply chain management design information systems-supported logistics processes (see section 5.4).

2.  Computer Aided Selling (CAS) organizers are familiar with the automation of the sales area (see section 5.1.2).

3.  Internet systems designers conceive portals and solutions for Internet commerce.

4.  Information brokers answer information queries through the use of the Internet and external databases.

For the operation of information systems one can identify many professions that require an information systems or an electrical engineering background. For business information systems specialists the following are of interest:

1.  *Database administrators* design and implement the structure of data that are stored in central databases. Partially they may also take on tasks pertaining to the protection and security of data.

2.  *Help desk staff* offers support for the functional departments during the selection and deployment of PCs and workstations, as well as for the needed software packages (see section 7.2.3).

The *mixed jobs* category is information systems-oriented and the information systems-dependent tasks (and knowledge) are distributed about equally.

In departments within which information systems support plays an important role, such as production planning and control (PPC systems), controlling (e.g., target/actual comparisons, special invoices) or sales (e.g., proposal quotation systems) *information systems coordinators* are in demand. This is somewhat of a liaison position in that this individual speaks the language of the users, as well as the one of the information systems specialists. In that role they have the opportunity to contribute significantly to the further development of the information systems penetration of functions and processes in his/her area. The information systems coordinator is especially then important when the responsibility for software development projects are located rather with the functional departments (in Fig. 7.2.3/2 right). *Fringe jobs*, on the other hand, are of a nature in which the contact to information systems merely plays a peripheral role. Individuals within this group use application systems or the available information on the Internet. It follows that in the meantime at least a basic knowledge about information systems has to exist here as well. In that sense it should be known, e.g., for which tasks information systems are suited and how or where, respectively, operational processes may be carried out more efficiently based on information systems deployment. In the area of media management and archiving, e.g., this in turn are also conceptual specialists for media and information services.

## 7.4 Literature for Chapter 7

Benjamin/Wigand 95 — Benjamin, K., Wigand, R. T., Electronic Markets and Virtual Value Chains on the Information Superhighway. Sloan Management Review, 36 (1995) 2, pp. 62-72.

Dixon 95 — Dixon, T., Invisible Eyes - Report on Video Surveillance in the Workplace, No. 67, September 1995, http://www.austlii.edu.au/au/other/privacy/video/notes.html #101.

Fink 99 — Fink, B., Ausrichtung der IT auf die globalen Aufgaben eines multinationalen Konzerns, WIRTSCHAFTS-INORMATIK 41 (1999) 4, pp. 348-357.

Hess 96 — Hess, T., Entwurf betrieblicher Prozesse, Wiesbaden, Germany, 1996.

Mertens 00 — Mertens, P., Integrierte Informationsverarbeitung 1, Administrations- und Dispositionssysteme in der Industrie, 12th edition, Wiesbaden 2000.

Mertens/Knolmayer 98 — Mertens, P., Knolmayer, G., Organisation der Informationsverarbeitung, 3rd edition, Wiesbaden 1998.

Mowshowitz 02 — Mowshowitz, A.,Virtual organization : toward a theory of societal transformation stimulated by information technology – Westport, CT ; London, 2002.

Porter 99 — Porter, M. E., Competitive Advantage: Creating and Sustaining Superior Performance, 2nd edition, Boston 1998.

Reichmann 97 — Reichmann, T., Controlling : concepts of management control, controllership, and ratios, Berlin et al., Springer, 1997.

Schumann et al. 97 — Schumann, M., Itter, R., Müller, J., von Stegmann u. Stein, E., Informationsagenten zur online-basierten Entscheidungsunterstützung am Beispiel einer Kreditversicherung, IM Information Management & Consulting 12 (1997) 4, pp. 30-37.

Schumann/Hess 00 — Schumann, M., Hess, T., Grundfragen der Medienwirtschaft, 2nd edition, Berlin et al. 2002.

Strohmeyer 92

Strohmeyer, R., Die strategische Bedeutung des elektronischen Datenaustausches, dargestellt am Beispiel von VEBA-Wohnen, Zeitschrift für betriebswirtschaftliche Forschung 44 (1992) 5, pp. 462-475.

Tamm/Günther 00

Tamm, G.; Günther, O., Business Models for ASP Marketplaces, in: Hansen, H.R., Bichler, M.; Mahrer, H. (eds): ECIS 2000, Vol. 2, Vienna, pp. 968-975.

Wigand 97

Wigand, R. T., Electronic Commerce: Definition, Theory and Context. The Information Society, 13 (1997) 3, pp. 1-16.

Zerdick et al. 00

Zerdick, A., Picot, A., Schrape, K., Artopé, A., Goldhammer, K., Lange, U. T., Vierkant, T., López-Escobar, E., Silverstone, R., E-conomics: strategies for the digital marketplace, Berlin et al., 2000.

# Further Readings

## Basic Readings

Applegate, L. M, Austin, R. D. and McFarlan, F. Warren, Corporate Information Strategy and Management: Text and Cases, 6th edition, New York McGraw-Hill, 2003.

Gordon, S., Information Systems: A Management Approach, New York, Wiley, 2003.

Laudon, K. C. and Laudon, J. P., Management Information Systems: Managing the Digital Firm, 7th edition, Upper Saddle River, NJ: Prentice-Hall, 2001.

McLeod, R., Jr. and Schell, G. P., Management Information Systems, 8th edition, Upper Saddle River, NJ: Prentice-Hall, 2001.

McNurlin, B. C. and Sprague, R. H., Information Systems in Practice, 5th edition, Upper Saddle River, NJ: Prentice-Hall, 2003.

Turban, E., McLean, E. and Wetherbe, J., Information Technology for Management, 3rd edition, New York, Wiley, 2002.

Turban, E., Rainer, R. K. and Potter, R., Introduction to Information Technology, 2nd edition, New York, Wiley, 2003.

Upton, D., Designing, Managing, and Improving Operations, Upper Saddle River, NJ: Prentice-Hall, 1998.

## Chapter 2

Bengel, G., Verteilte Systeme, Client server computing für Studenten und Praktiker, Braunschweig, 2000.

Bidgoli, H., (Editor-in-Chief), Encyclopedia of Information Systems, Academic Press Publishing Company, San Diego California, 2002.

Kauffels, F.J., Network management problems, standards and strategies, Addison-Wesley, 1992.

Lemay, L., Java 2 in 21 days, Macmillan Computer Publication, 1999.

Messerschmitt, D.G., Networked Applications, A Guide to the New Computing Infrastructure, San Francisco, 1999.

Stansifer, R, The study of Programming Languages, Prentice-Hall, 1994.

Tanenbaum, A.S., Woodhull, A.S., Operating Systems Design and Implementation, 2nd edition, Prentice-Hall, 2000.

## Internet Resources

Competence Center XML: http://www.xml-network.de

Competence Network IT-Standardization: http://www.it-standards.de

PC Webopaedia: http://www.pcwebopaedia.com/

Whatis.com: http://whatis.techtarget.com/

## Chapter 3

Allen, S., Data Modeling For Everyone, Curlingstones, 2002.

Date, C.J., An Introduction to Database Systems, 6th edition, Reading/Mass., 1995.

Elmasri, R. A., Navathe, S. B., Fundamentals of Database Systems, 3rd edition, Boston et al., 1999.

Harrington, J. L., Object-Oriented Database Design Clearly Explained, San Diego et al., 1999.

Mertens, P. und Griese, J., Integrierte Informationsverarbeitung 2, Planungs- und Kontrollsysteme in der Industrie, 8th edition, Wiesbaden, Germany 2000.

Scheer, A.-W., Business Process Engineering – Reference Models for Industrial Companies, Berlin, Germany, 1994.

Taylor, D.A., Object Technology: A Manager's Guide, Boston et al.,1997.

Ullmann, J.D., Principles of Database and Knowledge-Base Systems, Volume I, 8th edition, Rockville, 1995.

## Chapter 4

Jackson, P., Introduction to Expert Systems, 3rd edition, Boston, 1999.

Mertens, P. und Griese, J., Integrierte Informationsverarbeitung 2, Planungs- und Kontrollsysteme in der Industrie [Integrated Information Processing 2: Planning and Control Systems], 9th edition, Wiesbaden, Germany 2002.

Mertens, P., Integrierte Informationsverarbeitung 1, Administrations- und Dispositionssysteme in der Industrie [Integrated Information Processing 1: Operative Systems], 13th edition, Wiesbaden, Germany 2001.

Scheer, A.-W., Business Process Engineering, Reference Models for Industrial Enterprises, 2nd edition, Berlin et al., Germany 1994.

Thierauf, R.J., Executive Information Systems: A Guide for Senior Management and MIS Professionals, Westport, 1991.

# Chapter 5

Adelsberger, H.H. and Kanet, J.J., The „LEITSTAND" – A new tool for computer-integrated manufacturing, Production and Inventory Management Journal, 32/1 (1991), pp. 43-48.

Banks, E., e-Finance: The Electronic Revolution, John Wiley & Sons, 2001.

Callon, J., Competitive Advantage Through Information Technology, McGraw Hill, 1996.

Gosling, P., Changing Money: How the Digital Age is Transforming Financial Services, Capital Books Inc., 2000.

Grönroos, Ch., Service Management and Marketing. A Customer Relationship Management Approach, John Wiley & Sons, 2000.

Heskett, J. L. et al., Service Breakthroughs: Changing the Rules of the Game, Free Press, 1990.

Klein, S. (Ed.), Information and Communication Technologies in Tourism, Springer Verlag, 1996.

Knolmeyer, G., Mertens, P., and Zeier, A., Supply Chain Management Based on SAP Systems – Order Management in Manufacturing Companies, Berlin et al., Germany, 2002.

Kondratieff, N., Long Wave Cycle, Penguin USA, 1984.

Mertens, P. und Griese, J., Integrierte Informationsverarbeitung 2, Planungs- und Kontrollsysteme in der Industrie [Integrated Information Processing 2: Planning and Control Systems], 9th edition, Wiesbaden, Germany 2002.

Mertens, P., Borkowski, V. und Geis, W., Betriebliche Expertensystem-Anwendungen [Business Expert Systems Applications], 3rd edition, Berlin et al., Germany 1993.

Mertens, P., Integrierte Informationsverarbeitung 1, Administrations- und Dispositionssysteme in der Industrie [Integrated Information Processing 1: Operative Systems], 13th edition, Wiesbaden, Germany, 2001.

Orlicky, J and Plossl, G.W., Material Requirements Planning, 2nd edition, New York et al., 1994.

Scheer, A.-W., CIM – Computer Integrated Manufacturing. Towards the Factory of the Future, 3rd edition, Berlin et al., Germany, 1994.

Stadtler, H. and Kilger, C. (eds.), Supply Chain Management and Advanced Planning – Concepts, Models, Software and Case Studies, Berlin, Germany, 2000.

Toomey, J.W., MRP II – Planning for Manufacturing Excellence, New York et al., 1996.

Turban, E. et al., Electronic Commerce – A Managerial Perspective, Prentice Hall, 2002.

Vollmann, T.E., Berry, W.L. and Whybark, D.C., Manufacturing Planning and Control Systems, 4th edition, Columbus, 1997.

Wiendahl, H.-P., Load-Oriented Manufacturing Control, Berlin et al., Germany, 1995.

## Chapter 6

Appelrath, H.J., Ritter, J.,Sap R/3 Implementation: Methods and Tools, Springer, Berlin, Germany, 2000.

Blanchard, B.S., Fabrycky, Systems Engineering and Analysis, Prentice Hall, 3rd edition, New York, 1998.

Booch, G., Object-Oriented Analysis and Design with Applications, Addison-Wesley Pub Co; 2nd edition, Mass., 1994.

Booch, G., Jacobson, I., Rumbaugh, J., Rumbaugh, J., The Unified Modeling Language User Guide, Addison-Wesley Pub Co; 1st edition, Boston, 1998.

Date, C. J., An Introduction to Database Systems, Addison-Wesley Pub Co, 7th edition, Mass. 2000.

Davis, R., Business Process Modelling With Aris: A Practical Guide, Springer, London, 2001.

Dennis, A, Wixom, B., Tegarden, D., Systems Analysis and Design: An Object-Oriented Approach with UML, John Wiley & Sons, New York, 2001.

Florac, W.A., Carleton, A.D., Measuring the Software Process, Addison-Wesley Pub Co, 1st edition, Mass.,1999.

Fowler, M., Rice,D., Foemmel, M., Hieatt, E., Mee, R., Stafford, R., Patterns of Enterprise Application Architecture, Addison Wesley Professional, 1st edition, London, 2002.

Futrell, R. T., Shafer D. F., Shafer, L. I., Quality Software Project Management, Prentice Hall; 1st edition, New York, 2002.

Lewis, W.E., Software Testing and Continuous Quality Improvement, CRC Press, 2nd edition, Auerbach, 2000.

Martin, J., Odell, JJ., Object-Oriented Methods: A Foundation, UML Edition, Prentice Hall PTR, 2nd edition, New York, 1997.

Moriguchi, S., Software Excellence: A Total Quality Management Guide, Productivity Press, New York, 1997.

Murch, R., Project Management : Best Practices for IT Professionals, Prentice Hall PTR, 1st edition, New York, 2000.

Pressman, R. S., Software Engineering: A Practitioner's Approach, McGraw-Hill Science/Engineering/Math, 5th edition, Boston, 2001.

Raskin, J., The Humane Interface: New Directions for Designing Interactive Systems, Addison-Wesley Pub Co, 1st edition, Boston, 2000.

Rob, P., Coronel, C., Database Systems: Design, Implementation, and Management, Course Technology, 5th edition, Cambridge, 2001.

Scholz-Reiter, B., Stickel, E. (Eds.), Business Process Modelling, Springer, Berlin, Germany, 1996.

Schumann, M., Schüle, H., Schumann, U., Entwicklung von Anwendungssystemen, Grundzüge eines werkzeuggestützten Vorgehens, Berlin, Germany, 1994.

Sommerville, I., Software Engineering, Addison-Wesley Pub Co; 6th edition, Harlow, 2000.

Sharp, A., McDermott, P., Workflow Modeling: Tools for Process Improvement and Application Development, Artech House; 1st edition, Boston, 2001.

Webber, A.B., Modern Programming Languages: A Practical Introduction, Franklin Beedle & Assoc, Wilsonville, 2002.

Whitehead, K., Component-Based Development: Principles and Planning for Business Systems, Addison Wesley Professional, 1st edition, London, 2002.

# Chapter 7

Boddy, D., Boonstra, A., Kennedy, G., Managing information systems: an organisational perspective, Harlow et al.: Financial Times/Prentice Hall, 2002.

Galliers, R., Leidner, D., Dorothy E., Strategic information management: challenges and strategies in managing information systems, 3rd edition, Oxford, Butterworth-Heinemann, 2002.

van Grembergen, W. (Ed.), Information systems evaluation management, presented at: The Information Resources Management Association International Conference, Hershey et al., IRM.

Haag, S., Cummings, M., McCubbrey, D. J., Management information systems for the information age, 3rd edition, Boston, Mass., McGraw-Hill, 2002.

Hoffer, J.A., George, J.F., Valacich, J.S., Modern Systems Analysis and Design, 3rd edition, Prentice Hall, 2002.

Hussain, D.S., Hussain, K.M., Information management:organization, management and control of computer processing, 3rd edition, New York et al., Prentice Hall, 1993.

Laudon, K.C., Laudon, J.P., Management information systems: managing the digital firm, 7th edition, Upper Saddle River, NJ, Prentice-Hall, 2002.

Maggs, P.B., Soma, J.T., Sprowl J.A., Internet and Computer Law: Cases-Comments-Questions (American Casebook Series and Other Coursebooks), Florence, KY.

McNurlin, B.C., Sprague, R.H., Information systems management in practice, 5th edition, Upper Saddle River, NJ, Prentice Hall, 2002.

Post, G.V., Anderson, D.L., Management information systems: solving business problems with information technology, 3rd edition, Boston et al., McGraw-Hill/Irwin, 2003.

Reichmann, T., Controlling: concepts of management control, controllership, and ratios, Berlin et al., Germany, Springer, 1997.

Scheer, A.-W., Principles of Efficient Information Management, 2nd edition, Springer Verlag, Berlin et al., Germany, 1991.

Taylor, A.G., Englewood, C., The organization of information, Wetsport, CT, Libraries Unlimited, 1999.

Ward, J., Peppard J., Strategic planning for information systems, 3rd edition, Chichester et al., Wiley, 2002.

Wigand, R., Picot, A., Reichwald, R., Information, organization and management: expanding markets and corporate boundaries. - Chichester et al. : Wiley, 1997.

Zerdick, A.; Picot, A.; Schrape, K.; Artopé, A.; Vierkant, K. E., Lopéz-Escobar, Silverstone, R., E-conomics : strategies for the digital marketplace, Berlin et al. : Springer, 2000.

# Index

# Extremely simplified functional model of an industrial firm

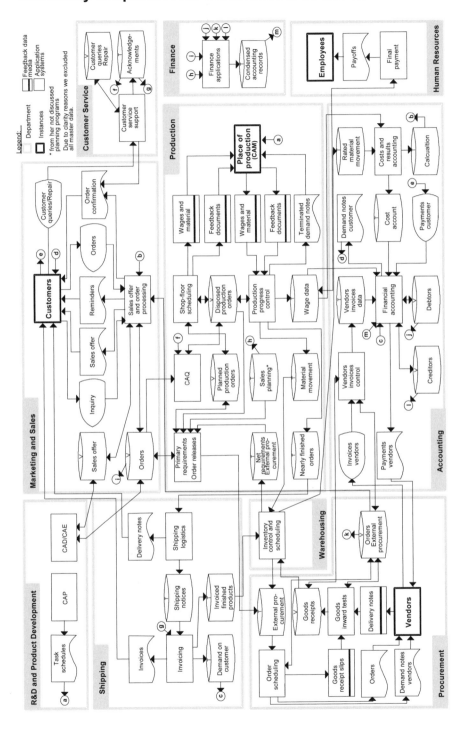